Hydraulics and Hydraulic Circuits

Dr.Ilango Sivaraman
(Acting Deputy Dean,Global College of Engineering& Technology, Sultanate of Oman)

COPYRIGHTS

All rights for this book are with the author Dr. Ilango Sivaraman.

© Dr. ilango Sivaraman 2016.

Book by the same author:

Pneumatics and Pneumatic circuits available on Kindle platform

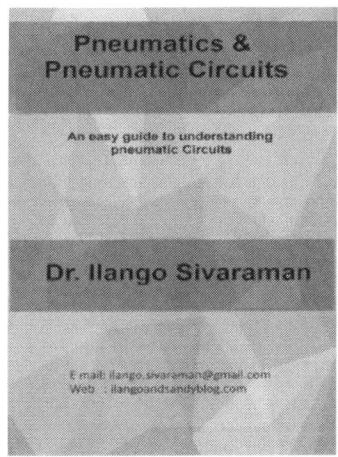

Table of Contents

COPYRIGHTS 7

1. FLUID POWER SYSTEMS 8
    1.1 LEARNING OUTCOMES 8
    1.2 TERMINOLOGY OF FLUIDS 8
    1.3 COMPARISON OF SYSTEMS 9
    1.4 TYPES OF HYDRAULIC SYSTEMS 9
    1.5 POINTS TO RECALL 11

2. UNITS IN HYDRAULIC SYSTEMS 12

3. FLOW OF HYDRAULIC FLUIDS 19
    3.1 LEARNING OUTCOMES 19
    3.2 FLUID FLOW IN HYDRAULIC SYSTEM 19
    3.3 LAMINAR AND TURBULENT FLOW 19
    3.4 BERNOULLI'S PRINCIPLE 20
    3.5 REYNOLDS NUMBER 21
    3.6 DARCEY-WEISBACH FORMULA 22
    3.7 POINTS TO RECALL 23

4. HYDRAULIC FLUIDS 24
    4.1 LEARNING OUTCOMES 24
    4.2 THE CHOICE OF HYDRAULIC FLUIDS 24
    4.3 HYDRAULIC OIL CHARACTERISTICS 26
    4.4 TYPES OF OIL GENERALLY USED IN HYDRAULIC SYSTEMS 26
    4.5 POINTS TO RECALL 28

5. AN OVERVIEW OF A HYDRAULIC SYSTEM 29
    5.1 LEARNING OUTCOMES 29
    5.2 HYDRAULIC SYSTEM SET UP 29
    5.3 HYDRAULIC SYSTEM CIRCUIT LAYOUT 29
    5.4 POINTS TO RECALL 31

6. HYDRAULIC PUMPS 32
    6.1 LEARNING OUTCOMES 32
    6.2 TYPE OF PUMPS 32
    6.3 PUMP PARAMETERS 34
    6.4 PUMP CLASSIFICATION 36
    6.5 RECIPROCATING PUMP – OPERATING PRINCIPLE 36
    6.6 ROTARY PUMPS 37

- 6.6.1 External gear pumps...37
- 6.6.2 Lobe pump...39
- 6.6.3 Internal gear pumps...41
- 6.6.4 Vane pump..42

6.7 PISTON PUMPS..47
- 6.7.1 Radial Piston pumps...47
- 6.7.2 Axial Piston pumps...49
- 6.7.3 Bent axis pump...52

6.8 USAGE OF PUMPS IN OPEN AND CLOSE HYDRAULIC SYSTEMS..........53
- 6.8.1 Closed Loop Hydraulic system...53
- 6.8.2 Open loop Hydraulic system..54

6.9 POINTS TO RECALL..54

7. HYDRAULIC ACTUATORS...56

7.1 LEARNING OUTCOMES:...56

7.2 CLASSIFICATION OF HYDRAULIC ACTUATORS...........................56
- 7.2.1 Hydraulic cylinders..56
- 7.2.2 Working Principle of Hydraulic cylinders...................................57
- 7.2.3 Specifications of Hydraulic cylinders...58
- 7.2.4 Symbols of Hydraulic cylinders for hydraulic circuit drawing.....58
 59
- 7.2.5 Cushioned and non-cushioned cylinders...................................59
- 7.2.6 Ram Cylinder...63
- 7.2.7 Telescopic cylinder..64

7.3 POINTS TO RECALL..65

8. HYDRAULUC ACTUATORS-2...66

8.1 LEARNING OUTCOMES..66

8.2 HYDRAULIC MOTORS..66

8.3 HYDRAULIC MOTOR COMPARISON WITH ELECTRIC AND PNEUMATIC MOTORS..66

8.4 TYPES OF HYDRAULIC MOTORS AND SELECTION OF MOTORS....67

8.5 POPULAR TYPES OF HYDRAULIC MOTORS.................................68
- 8.51 Gear Motor..68
- 8.5.2 Vane Motor...69
- 8.5.3 Piston motors...71

8.6 DISADVANTAGES OF HYDRAULIC MOTORS................................71

8.7 HYDRAULIC CIRCUIT SYMBOLS FOR HYDRAULIC MOTORS.........72

8.8 POINTS TO RECALL..72

9. HYDRAULIC ELEMENTS..73

9.1 LEARNING OUTCOMES..73

9.2 HYDRAULIC ELEMENTS AND ACCESSORIES..............................73
- 9.2.1 Hydraulic Control elements...73
- 9.2.2 Hydraulic accessories..73

9.3 POINTS TO RECALL..74

10. UNDERSTANDING HYDRAULIC CIRCUITS 75

10.1 LEARNING OUTCOMES 75
10.2 FIRST STEP IN THE PROCESS OF BUILDING A HYDRAULIC CIRCUIT 75
10.2.1 System pressure relief Valve 76
10.2.2 Pressure Reducing valve 79
10.2.3 Pressure sequence valve 81
10.3 POINTS TO RECALL 82

11. UNDERSTANDING HYDRAULIC CIRCUITS-2 83

11.1 LEARNING OUTCOMES 83
11.2 SIZING OF HYDRAULIC CYLINDER, PUMP AND THE MOTOR FOR HYDRAULIC CIRCUIT 83
11.3 FUNCTION OF DIRECTION CONTROL VALVE 86
11.4 A SIMPLE HYDRAULIC CIRCUIT -CONNECTING DIRECTION CONTROL VALVE TO THE CYLINDER 87
11.5 POINTS TO RECALL 89

12. TYPES OF DIRECTION CONTROLVALVES AND THEIR ACTUATION 90

12.1 LEARNING OUT COMES 90
12.2 DIRECTION CONTROL VALVE CLASSIFICATIONS 90
12.3 PILOT OPERATED DC VALVE 94
12.4 CHECK VALVE 96
12.5 PILOT OPERATED CHECK VALVE 99
12.6 POINTS TO RECALL 100

13. UNDERSTANDING HYDRAULIC CIRCUITS-3 101

13.1 LEARNING OUTCOMES 101
13.2 PRESSURE REDUCING VALVE & PRESSURE REDUCING VALVES IN A HYDRAULIC CIRCUIT 101
13.3 PRESSURE SEQUENCE VALVE IN A HYDRAULIC CIRCUIT 103
13.4 HYDRAULIC MOTORS IN CIRCUITS 105
13.4.1 Over center valve with a hydraulic motor 106
13.5 HYDRAULIC REGENERATIVE CIRCUIT 109
13.6 POINTS TO RECALL 112

14. FLOW CONTROL VALVES 113

14.1 LEARNING OUTCOMES 113
14.2 FUNCTION OF FLOW CONTROL VALVES 113
14.3 TYPES OF FLOW CONTROL VALVES 113
14.4 POINTS TO RECALL 117

15. UNDERSTANDING HYDRAULIC CIRCUITS-4 118

15.1 LEARNING OUTCOMES 118
15.2 FLOW CONTROL VALVE USING METER IN METHOD 118

15.3 METER OUT CIRCUIT ... 120
15.4 BLEED OFF CIRCUITS ... 122
15.5 SPEED AND FEED CONTROL IN ONE DIRECTION- MACHINE TOOL APPLICATIONS ... 123
15.5.1 Speed and feed control – Meter in circuit ... 123
15.5.2 Speed and feed control – Meter out circuit ... 125
15.5.3 Speed and feed control in both directions of piston movement ... 126
15.5.4 Standard manifold for dual speed ... 127
15.6 HYDRAULIC CIRCUIT FOR PRESS APPLICATION USING DOUBLE PUMP ... 128
15.7 POINTS TO RECALL ... 135

16. ACCUMULATOR APPLICATIONS ... 136
16.1 LEARNING OUTCOMES ... 136
16.2 FUNCTIONS, TYPES AND SYMBOLS OF HYDRAULIC ACCUMULATORS ... 136
16.3 BLADDER TYPE ACCUMULATORS ... 137
16.4 SIZING -ACCUMULATORS ... 139
16.5 ACCUMULATOR CIRCUITS ... 139
16.6 LARGE FLOW RATE REQUIRED FOR SHORT PERIODS – APPLICATION IN PLASTIC INJECTION MACHINES ... 140
16.6.1 Accumulator unloading valve ... 142
16.7 FEW OTHER APPLICATIONS OF ACCUMULATOR ... 144
16.8 POINTS TO RECALL ... 145

17. HYDRAULIC CIRCUITS IN STACKABLE FORM ... 146
17.1 LEARNING OUTCOMES ... 146
17.2 STACKABLE VALVES ... 146
17.3 POINTS TO RECALL ... 149

18. INTRODUCTION TO PROPORTIONAL VALVES ... 150
18.1 LEARNING OUTCOMES ... 150
18.2 PROPORTIONAL CONTROL VALVES ... 150
18.4 PROPORTIONAL SOLENOIDS ... 151
18.5 PROPORTIONAL CONTROL VALVES ... 152
18.6 POINTS TO RECALL ... 154

19. SERVO HYDRAULICS ... 155
19.1 LEARNING OUTCOMES ... 155
19.2 WHAT IS SERVO HYDRAULICS ... 155
19.3 COMPARISON OF PROPORTIONAL HYDRAULICS WITH SERVO HYDRAULICS ... 155
19.4 ELECTRO HYDRAULIC SERVO VALVES ... 156
19.4.1 Servo Valves construction ... 157

19.5 POINTS TO RECALL 158

20. LOGIC OR CARTRIDGE VALVES 159

20.1 LEARNING OUTCOMES 159

20.2 WHAT IS A LOGIC/CARTRIDGE VALVE? 159

20.3 ADVANTAGES OF CARTRIDGE VALVES 159

20.4 CONSTRUCTIONAL FEATURES OF POPPET TYPE SLIP IN CARTRIDGE VALVES 159

20.5 POINTS TO RECALL 162

21. APPENDIX 1 - HYDRAULIC TERMINOLOGY 163

22. APPENDIX2-USEFUL FORMULAE(FPS) 167

1. Flow velocity 167

2. Hydraulic motors 167

23. APPENDIX3-BIBILIOGRAPHY 168

1. FLUID POWER SYSTEMS

1.1 LEARNING OUTCOMES

On completion of this chapter the student will understand:

1. The meaning of the term Fluid and its applications.
2. The types of fluid power systems.

1.2 TERMINOLOGY OF FLUIDS

The state of the matter is divided into Solids and fluids. The easiest way to remember the meaning of 'fluid' is that it 'flows' and has no shape unlike Solids. It generally assumes the shape of its container.

Fluids are further divided into liquids and gases (including air). If we take compressed air, it is a fluid and dealt under the name 'Pneumatics.' Liquids like water and oil are dealt under the caption of 'Hydraulics. However, the term Hydra refers mainly to water and hence to add clarity to the subject where mineral oil is used in machines, it is not uncommon to refer the application as 'Oil hydraulics.'

We should now look at the term Fluid power. Most of us are clear about electrical power (Household applications like lighting/air conditioning and other household applications.). Many of us are also clear about mechanical power – Engines, Gear boxes etc.,

Fluid power - Fluid power is the use of fluids under pressure to generate, control, and transmit power. The fluid can be compressed air- Pneumatics. (there are ever so many industrial applications of compressed air – the one most common application that we come across every day in city life is the air brake applications in commercial vehicles like buses and lorries. - if you ever changed your car tire in a well-equipped shop, you would have seen a car tire changer entirely using Pneumatics. Fluid power can be also be done by using water or mineral oil. There are many advantages of using mineral oil over any other liquid and we shall see these details in the following pages.

It should be remembered that fluid power does not mean that it is exclusively fluid power– Generally it is a mix of all the three forms of power (Electrical, Mechanical and Fluid power), so that over all we optimize on the efficiency of the machine used.

The above statement can be supported by citing the example of off highway vehicles like tippers/dumpers and shovels etc., where, we use electrical batteries and starters for getting electrical power, Engines, transmission systems for Mechanical and the operation of tipping/dumping mechanism using hydraulic power.

1.3 COMPARISON OF SYSTEMS

The three systems – Hydraulics, Pneumatics and Electrical systems are compared in table 1 given below.

S.No	Factor	Hydraulics	Pneumatics	Electrical
1	Output- Linear motion	Simple by using Cylinders	Simple by using Cylinders	Difficult.
2	Output- Rotary motion	Simple- by using Hydraulic motors	Simple – but normally for high speeds	Simple and normally most popular
3	Energy output transmission/ distribution.	Done using hoses and piping. But limited to about 100 meters (328 ')	Done using hoses and piping. Limited to about 1000 meters (3280')	Unlimited distances – done using cables.
4.	Energy storage	Limited- using Accumulators	Easy – using Reservoir.	Difficult – using batteries.
5.	Energy cost	Moderate to high	Moderate	Low
6.	Problems	Leakage/ High pressure	Leakage/ Low pressure	Electrical short circuits

Comparison of Electrical/ Pneumatic and Hydraulic systems

It is possible to ask a question – "What are the advantages of a hydraulic systems?". The answer is that each system has their own merits and demerits. Only based on the application, we can make our choice for an appropriate system.

1.4 TYPES OF HYDRAULIC SYSTEMS

Hydraulic systems are categorized in to two types. They are hydro static systems and Hydro dynamic systems.

In hydro static systems – The fluid is kept in a confined space. The fluid is then pumped to move in the confined space. As the fluid is moving in its restricted confined pipes, it encounters a load (say, a load is to be pushed). The liquid is constrained to move against the load and the fluid does this by exerting pressure on the load. Hence the fluid pushes to move against the load pressure. The following illustration, for added clarity.

In this figure 1.1, Oil is pumped in to a pipe which is connected to a hydraulic motor. The hydraulic motor is rotating a heavy load. As the fluid encounters the vanes of the hydraulic motor, it has no other alternative but to push the vane against its load.

Why?

Because, more oil is being pumped into the pipe line. The hydraulic oil has a property of non-compressibility cannot be compressed. The only way is that as it is being pressurized to move, it transfers this pressure energy to the vanes of the motor and pushes it away. Subsequent vanes attached to the motor also get pushed by the oil and the motor starts rotating.

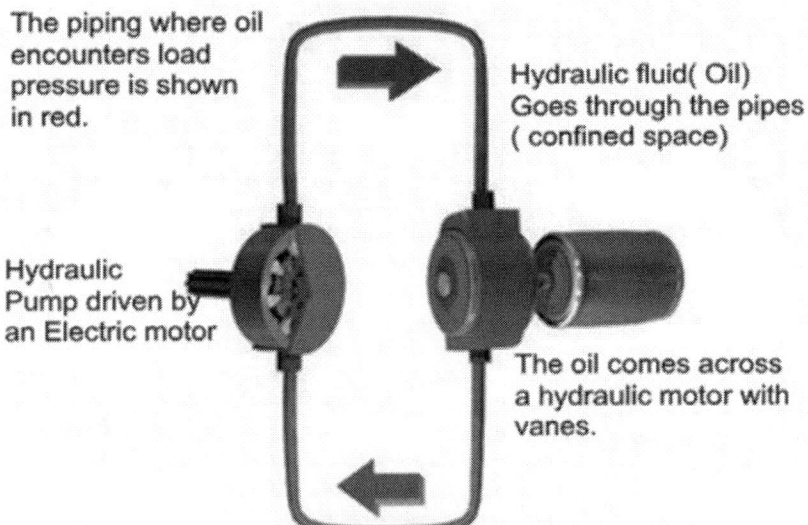

Figure 1 1 Hydrostatic application- application of Pressure energy on the motor vanes

In hydro dynamic systems, the power of transmission by the oil is by kinetic energy. The differences compared to hydrostatic system are:

1. The fluid is in motion.
2. The fluid is not pressurized
3. It is the kinetic energy of the flow that is transferred for tackling the load.

Figure 1.2, shows the example application of hydro dynamic system.

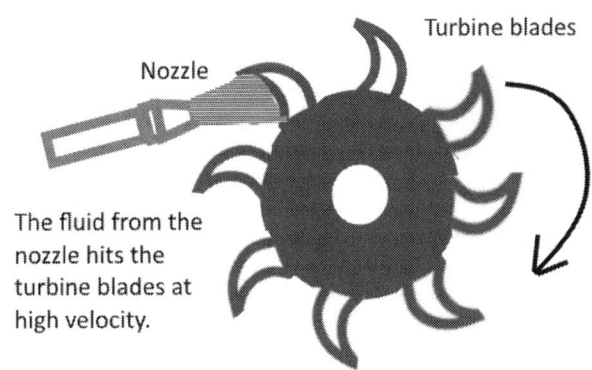

Figure 1.2 Application of Kinetic energy of the fluid

1.5 POINTS TO RECALL

1. A system for carrying out a work is normally made up considering all types of power systems mix for optimizing efficiency and power.

2. There are two kinds of hydraulic systems and we shall be considering mainly the hydro static systems in this book.

2. UNITS IN HYDRAULIC SYSTEMS

2.1 LEARNING OUTCOMES

By the end of the chapter the students will learn about:

1. Different units used in oil hydraulic systems.

2. Characteristics of a hydraulic system.

2.2 HYDROSTATIC PRESSURE

Hydrostatic pressure is the pressure that is exerted by a fluid at equilibrium at a given point within the fluid, due to the force of gravity. Hydrostatic pressure increases in proportion to depth measured from the surface because of the increasing weight of fluid exerting downward force from above.

The static fluid pressure at a given depth does not depend upon the total mass, surface area, or the geometry of the container.

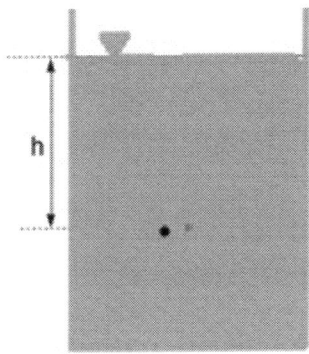

Figure 2 1 Hydrostatic pressure

Please refer Figure 2 1. At the particle P the pressure of the liquid over it is given by the following formula:

P Pressure = ρ g h

Where P is hydrostatic pressure given in Pascals (N/m2, lbf/ft2, psf)

ρ = Density of the liquid given in kg/m3 (slugs/ft3)

g = Acceleration due to gravity in m/s2 (9.81 m/s2, 32.17405 ft/s2)

h = level of the column of liquid above the particle in meters (feet)

The alternative system (FPS) of measurement is given in brackets.

There are a few principles that govern the Oil Hydraulic systems.

2.3 NON-COMPRESSIBLE HYDRAULIC OIL

All liquids are non-compressible. It follows that the fluid (oil) used in the hydraulic system, is non-compressible for all practical purposes. However, we can qualify the above statement by saying that the petroleum oil that will be used in hydraulic power pack is slightly compressible at pressures over 70 Bar(X 14.5 psi =1015 pounds per sq.in.). In general, for the systems we are going to study, we will treat the oil as non-compressible.

2.4 PRESSURE

In our hydraulic oil systems, we are going to use petroleum oil(Mineral oil) as the medium to do the work for us. The oil will be pumped in to a cylinder(For linear movement output) or to a hydraulic motor for rotary movement output.

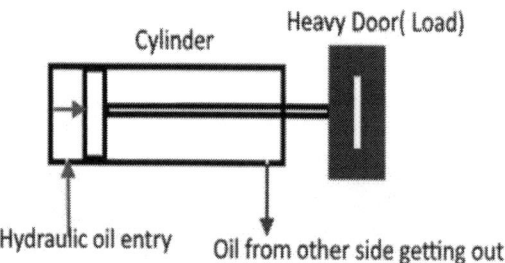

Figure 2 2 Pressure on the oil

For instance, let us take hydraulic oil pumped into a hydraulic cylinder to move a load. Please refer Figure 2 2. Hydraulic oil when pumped in gets into the cylinder as shown by the red arrow. As the oil gets in it acts on all sides of the enclosed space. The figure shows only one red arrow acting on the piston of the cylinder, but oil exerts pressure on all sides and at perpendicular direction to the wall of the enclosed space). As the oil continues to flow in and since it is a confined space, the pressure in the liquid increases and this pressure is transferred to the piston, which starts to move, pushing a heavy door attached to the piston rod.

In effect the pressure in the liquid increases as there is a resistance offered by the weight of the piston, piston rod and the heavy door. When the pressure is marginally more than the resistance, the door along with the piston is pushed away by the oil.

It means that the resistance offered by the load is the pressure felt by the liquid. If there is no load(No heavy door attached to the piston rod), then, the pressure required to overcome the self-weight is enough for the hydraulic oil to push or move/extend the piston assembly.

2.5 FLOW RATE

We will be coming across different sizes of pumps. The pump capacity has to be properly chosen. Prior to that, it is important to understand what is the effect of having a pump that delivers more oil in to the cylinder than the calculated pump size.

The effect is that the more quantity of oil is delivered, the faster the piston is moved. This means that the oil flow into the cylinder governs the speed of movement of the cylinder or the speed of a hydraulic motor.

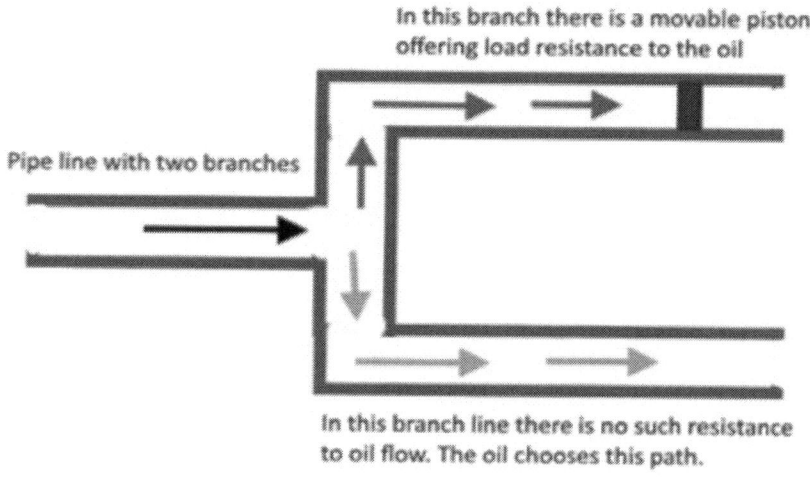

Figure 2 3 Oil chooses the path of least resistance

2.6 OIL CHOOSES A PATH OF LEAST RESISTANCE

In any hydraulic system, the oil chooses a path of least resistance. Let us imagine that the hydraulic engineer has provided two paths for the oil to flow in the pipe line, as in the Figure 2 3

The green arrows indicate the path of least resistance for the oil flow.

2.7 APPLICATION OF PASCAL'S LAW

Pascals law states that "Pressure applied on a contained fluid is transmitted undiminished in all the directions and acts with equal force and at right angles to them".

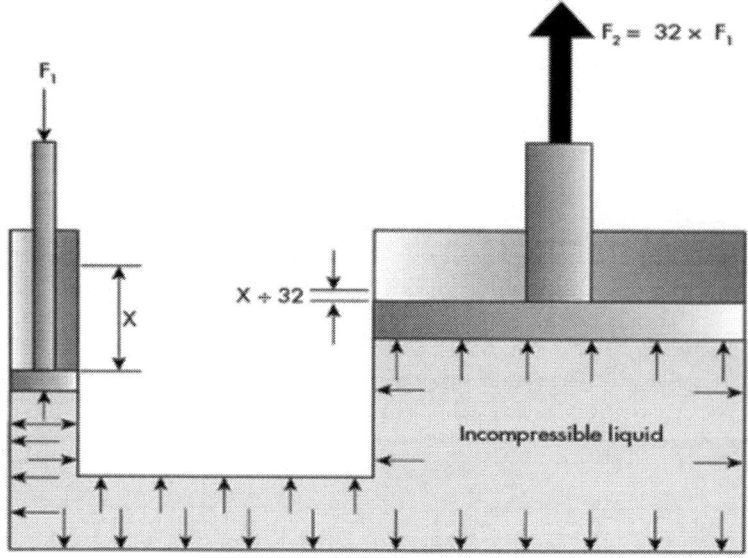

Figure 2 4 Application of Pascal's law

Please refer Figure 2 4. In the above figure, a small force F1 is applied on the small area a(say). Let F2 be the force felt on the larger end – on the larger piston which has an area A. Then,

F2 / f1 = A/a.(The student should be well aware that Pressure= Force/Area)

When you apply force f1 on to the smaller piston of area 'a', the piston exerts pressure on the contained liquid and this is transmitted undiminished in all the directions (as shown by the arrows acting in perpendicular directions.).

The distance travelled by the small piston is X. In view of the pressure transmitted by the contained liquid,the larger piston also is moved, although by a smaller distance, a fraction of X. Let us call this as ✲.The distance of travel of the pistons are inversely proportional to their areas.

X / ✲ = A/a; F/f = A/a = X/ ✲

If the area A is 32 times bigger than a, then,

F2 = 32 * f1

And ✲ = X / 32

Let us consider specific figures for the areas and forces.

If we apply a force of 50 Kg on the smaller piston of 50cm2 and if the area of the larger piston is 4 times(not 32 times(!), and is 200 cm2,then the larger force on the 200 cm2is worked out as follows.

F2/f1 = A/a

F2 = 50 Kg * (200 cm2 /'50 cm2)

= 200 Kg.

While on this topic, we must also know that

Pressure(Kg/cm2) = Force (Kg)/Area(cm2)

2.8 PRESSURE DROP

We will be hearing this term often in the practical applications of the hydraulic system.

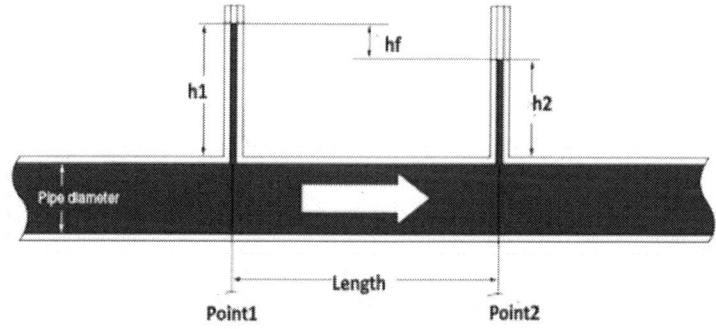

Figure 2 5 Pressure loss due to friction

It is possible that the pressure is lost when the oil flows through the pipe. The pressure loss occurs due to the pipe friction. Please refer Figure 2 5. At point 1, pressure is h1 and as the oil flows through the length of L, the pressure drops by hf and at point 2 it is less and measured as h2.

Following factors influence the amount of pressure drop.

1. Friction- The fluid rubbing against the inside of pipe or hose assembly.

2. Type of fluid- Viscous fluids tend to have more friction and will result in higher pressure drop.

3. Temperature of fluid- If the temperature of the fluid is more, the fluid thins and flows with more ease and the pressure drop is less. (Automotive engine oil gets heated up during usage.)

4. Length/ID(Inner Diameter) of the pipe/hoses: The Figure 2 5 showed the effect of

pressure drop over a particular length. If the length is more, the pressure drop will be more. If there are bends or other valves in the pipe/ hoses, then also the pressure drop will increase.

If the ID is more, the flow velocity will be less and the pressure drop also will be less.

5. Flow rate: if the flow rate increases then the pressure drop also will increase.

2.8.1 Effect of pressure drop in a hydraulic system

Let us imagine that your design engineer tells that to move a load by a hydraulic cylinder, a pressure of 100 Bar (1470psi) is required. In case the pressure drop of the oil till it enters the cylinder is 10 Bar (147 psi), then we should ensure that the system is capable of delivering 110 Bar (1617 psi) at the output end.

2.9 POINTS TO RECALL

1. The units of force, pressure and flow rate are useful in understanding the concepts.

2. Properties of hydrostatic systems like oil choosing the path of least resistance, load creating resistance to flow is pressure, pressure drop and application of Pascals law are important aspects to remember in respect of hydraulic systems.

3. FLOW OF HYDRAULIC FLUIDS

3.1 LEARNING OUTCOMES

By the end of this chapter, the student will understand:

1. The concept of flow of fluids in hydraulic systems.

2. Bernoulli's Principle and Reynolds number and their significances in Hydraulics.

3.2 FLUID FLOW IN HYDRAULIC SYSTEM

Normally in a hydraulic system, the output is the actuation of hydraulic cylinders/Hydraulic motors. That means, load is overcome by the linear movement of a hydraulic cylinder/ hydraulic motor. We need the flow of the hydraulic oil to actuate the hydraulic cylinder/ hydraulic motor.

What about pressure?

Pressure gives the actuator (Cylinder or motor) the force required but flow is essential to cause the movement.

While pressure is measured in Bar, kg/sq.cm or psi, measurement of flow can be done in two ways.

a) By measuring the velocity of the fluid- it is the average distance the fluid particles travel per unit of time and is measured in meters/sec (feet per second).

b) By measuring the flow rate- it is a measure of the volume of fluid passing a point in a given time – the units are l/min(gpm) or Cubic cm per minute(Cubic inches per minute)

3.3 LAMINAR AND TURBULENT FLOW

It is preferred that when fluid particles flow through a pipe, they move in straight, parallel flow paths. This is known as 'laminar flow'. With laminar flow the friction is minimum. Laminar flow occurs at low velocities and in straight piping or piping where the path change is gradual.

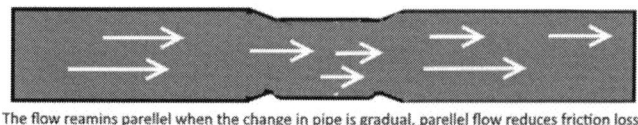

The flow reamins parellel when the change in pipe is gradual. parellel flow reduces friction loss

Figure 3 1 laminar flow

In 'Turbulent flow', the fluid particles do not move smoothly parallel to the flow direction. The turbulent flow is caused by abrupt changes in direction or cross section or by high velocity. This results in increased frictional loss, generates heat and increases operating pressure.

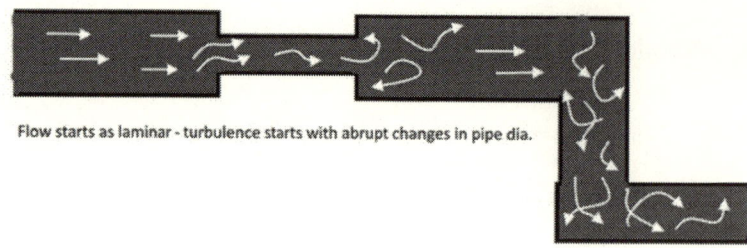

Figure 3 2 Turbulent flow

3.4 BERNOULLI'S PRINCIPLE

Hydraulic Fluid in a system contains energy in two forms.

Kinetic energy – By virtue of the fluid's weight and velocity, and,

Potential energy in the form of pressure.

Bernoulli's principle states that if the flow rate in a system is constant, the sums of Kinetic energy and the pressure energy at different points in a system must be constant. If the kinetic energy decreases, then the decrease will be compensated by an increase in potential energy. This happens when the pipe diameter increases resulting in a lesser velocity of the fluid.

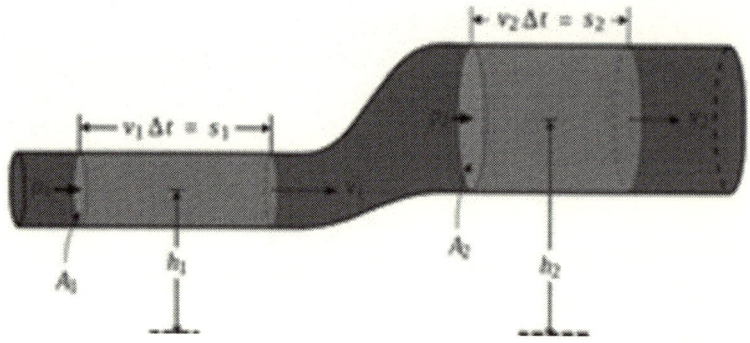

Figure 3 3 Bernoulli's Principle

Please refer Figure 3 3. The flow of fluid through a pipe that is smaller in section and also the larger section is shown.

P1 Pressure at point 1, h1 is the height from a common datum level, V1 is the velocity and S1 is the distance travelled in Δt. Corresponding notations for section 2 is also shown.

It can be shown, as per Bernoulli's principle that:

$$P_1/\rho g + (v_1^2)/2g + h_1 = P_2/\rho g + (V_2^2)/2g + h_2$$

That is Potential energy plus the kinetic energy (Total energy remains same at different locations of a hydraulic system. However, the equation is correct, only if we also consider the loss due to friction and add the same to the second point. Let us say it is hf

$$P_1/\rho g + (v_1^2)/2g + h_1 = P_2/\rho g + (V_2^2)/2g + h_2 + h_f$$

It is an important factor especially in the design of hydraulic valves, where the passage diameter varies and the maximum flow rate through the valve is to be designed or to be selected for use in a hydraulic system. Earlier, we have seen that the pressure is the resistance to flow. The valves designed/selected should offer the least resistance to flow (Minimum pressure drop).

The most popular application of Bernoulli's principle is in aerodynamics- in the design of aircraft wings.

3.5 REYNOLDS NUMBER

We have seen the laminar flow and the turbulent flow through pipes in a hydraulic system. We are aware that the laminar flow turns turbulent beyond a particular value of velocity. This velocity is known as critical velocity.

The specific calculation of the Reynolds number, and the values where laminar flow occurs, will depend on the geometry of the flow system and flow pattern. The common example is flow through a pipe, where the Reynolds number is defined as:

Re, Reynolds Number = $\rho V D_H/\mu = V D_H/v = Q D_H/v A$

D_H is the hydraulic diameter of the pipe; its characteristic travelled length, L, (m).

Q is the volumetric flow rate (m3/s).

A is the pipe's cross-sectional area (m2).

V is the mean velocity of the fluid (SI units: m/s).

µ is the dynamic viscosity of the fluid (Pa·s = N·s/m2 = kg/(m·s)).

v is the kinematic viscosity of the fluid,

ρ is the density of the fluid (kg/m3).

For such systems, laminar flow occurs when the Reynolds number is below a critical value of approximately 2,040, through the transition range is typically between 1,800 and 2,100 (Kerstin Avila1, 2011).

From this books perspective, it is enough if the student has understood that the laminar flow to turbulent flow happens beyond a particular value of Reynolds number.

3.6 DARCEY-WEISBACH FORMULA

We have also seen earlier that whenever there is a flow in a hydraulic piping, there is a pressure drop. This is because of the friction loss due to the length of the pipe as well as due to the number of valves and fittings (Bends, elbows and Tees) used in the piping. If more fittings are used the pressure drop will be more.

If the friction loss is to be expressed in pressure head, then,

hf = head loss due to friction = (4flv2)/ 2gd

f = coefficient of friction of the pipe used

l = length of the pipe in meters (feet)

v = velocity of the fluid meters/ sec (feet/sec)

d = diameter in meters(feet)

From the above formula, we can see that the loss in a pipe is proportional to the velocity head (v2/2g), length of the pipe and inversely proportional to the diameter of the pipe.

The above formula can be simplified for calculating the loss due to friction for valves and fittings as follows;

hfv = K [V2/2g]

hfv = Friction loss due to valve /fittings

K = Loss coefficient for a particular valve or fitting(Normally available from manufacturer of valves/fittings)

V = Flow velocity

In terms of flow rate through the pipe, the following formula is used.

Q = a * V

Q= flow rate in liters per minute (gallons per minute) [convert to cm3/min or ft3/ min]

a= area of the cross section of the pipe in cm2 (ft2)

V = Flow velocity centimeters/minute or feet/minute

3.7 POINTS TO RECALL

1. Understanding of. Laminar and Turbulent flow- its impact on hydraulic systems.

2. Bernoulis Principle, Reynold number and Darcey Weisbach formula and their significance in flow through pipes.

4. HYDRAULIC FLUIDS

4.1 LEARNING OUTCOMES

By the end of this chapter the student will understand:

1. the functions and characteristics of hydraulic fluids in oil hydraulics.
2. The different types of hydraulic oils used in the industry.

4.2 THE CHOICE OF HYDRAULIC FLUIDS

The hydraulic fluid used in hydraulic system is also referred as medium of the hydraulic system. The most common medium used in oil hydraulics is mineral oil. The types of mineral oil used is dealt at the later part of this chapter.

We must also examine why water is not a preferred medium in hydraulic systems. There are a few very common reasons why we do not prefer water in our hydraulics systems- though hydraulics mean water to most of the people.

1. Water leads to corrosion of other elements /parts used in the system.
2. Supports bacterial growth.
3. Easily evaporates and gets contaminated.
4. It has Poor lubricity
5. it is prone to leakage.

The choice of hydraulic medium should meet the following functional criteria.

1. Power transmission or energy transfer
2. Lubrication
3. Sealing
4. Cooling (heat transfer).

Power transmission function

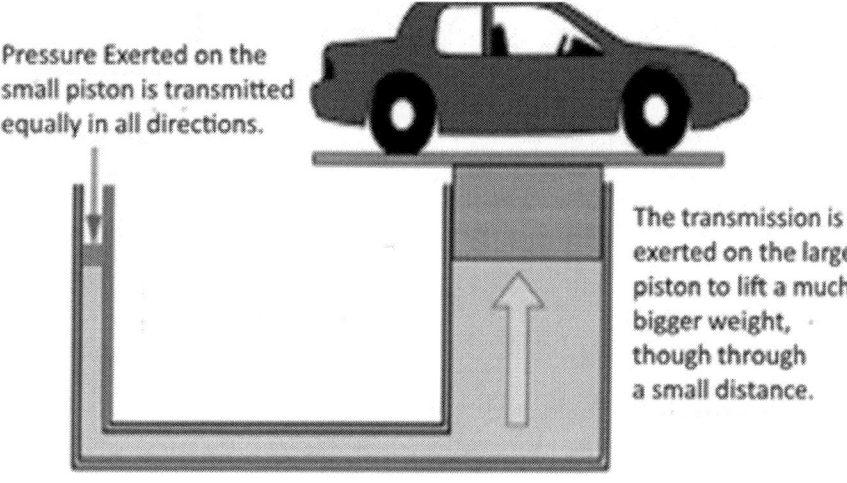

Figure 4 1 Power/ energy transfer

The Figure 4 1 , shows how the power is transferred/transmitted by the medium used.

Lubrication Function -The medium chosen should also function as a lubricator of internal parts of the hydraulic system.

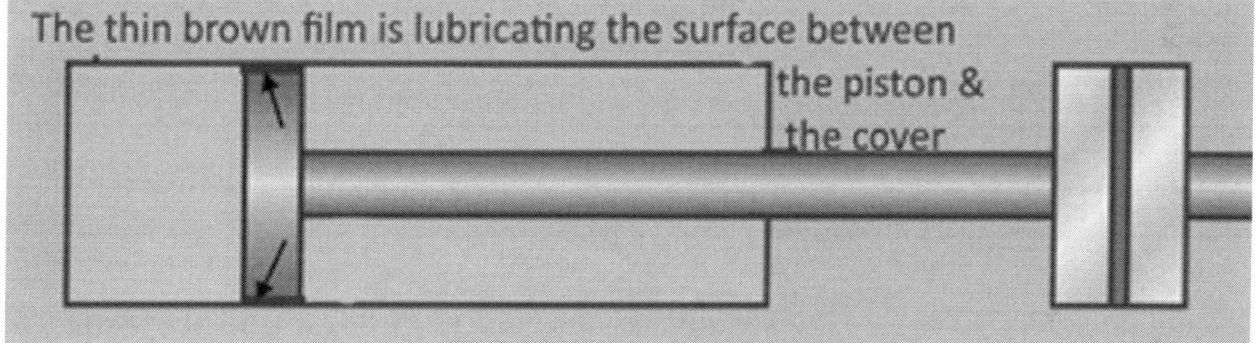

Figure 4 2 Internal lubrication

Please refer Figure 4 2 . The hydraulic oil not only used as a medium of transmitting power, it also serves as a lubricant of internal parts. In the above figure, you could see a thin brown line drawn to show how the hydraulic oil also lubricates the surface between the piston and the cylinder cover by providing a thin film of lubrication between the parts.

Sealing function-In the above figure, the medium of oil not only provides a film of lubrication as shown, the same film acts as a sealing between either side of the piston. It does not allow the oil from one side of the piston to cross over to the other side.

Cooling-The hydraulic oil should also be capable of absorbing the heat generated by the moving parts of the system- This would avoid the system operating at higher temperatures and the system performance would suffer, if the internal heat generated is not cooled.

4.3 HYDRAULIC OIL CHARACTERISTICS

Viscosity of oil - The viscosity of the medium (oil) chosen should be right for the system. Viscosity is the resistance to flow. If the viscosity is low, it would flow easily and result in leakage. If the viscosity is more, there will be increased pressure drop/slow operation/ increased power consumption.

It is recommended that the viscosity of the oil chosen is between 46 to 68. These numbers indicate the time in seconds for a particular oil to flow through a standard orifice and from a standard specified size of container.

Pour Point – The pour point is the lowest temperature at which a fluid will start to flow. Generally the pour point of the hydraulic oil chosen should be 200 F (approx. 70C) below the lowest temperature the system would encounter in its operation.

Oxidation resistance - Oxidation means burning of the oil itself. In other words, it is a chemical union of the oil with oxygen of the atmosphere. petroleum oils are particularly susceptible to oxidation, since oxygen readily combines with both carbon and hydrogen elements present in the oil.

Oxidation of the oil leads to sludge/muck /varnish formation which would greatly affect the performance of the system.

It is necessary to constantly monitor the hydraulic system to see that it is not getting heated up. If the operating temperature goes beyond 800C (800F), it is advisable to check what is causing this high temperature and provide a cooling method for lowering the temperature.

Oil companies use additives to the hydraulic oils to increase the oils resistance to oxidation.

Rust and Corrosion - Rust is the chemical union of oxygen with iron(steel). Corrosion is a chemical reaction between a metal and acid. Acids form because of water(moisture in the atmosphere and certain chemicals.

The intensity of the problem is again reduced by the oil companies mixing certain additives to the oil.

Demulsibility - This is the property of the fluid that decides how the oil will behave when mixed with water. The medium (hydraulic oil) chosen should have high degree of demulsibility – That is it does not mix with water.

4.4 TYPES OF OIL GENERALLY USED IN HYDRAULIC SYSTEMS

In most cases, the type of hydraulic oil used is hydraulic mineral oil which is a byproduct when crude oil is refined. Following are the most commonly used hydraulic mineral oils used.

Three common varieties of hydraulic fluids found on the market today are petroleum-based, water-based and synthetics.

Petroleum based mineral hydraulic oils-The classifications of hydraulic oil are a subgroup of different fluids with varying performance levels. Below is a list of common hydraulic oil classifications and their brief descriptions:

Mineral hydraulic oils are classified in accordance with ISO 6743/4 and DIN 51524.

Description	ISO	DIN
Mineral oil without additives	HH	H
Type HH + oxidation and corrosion-inhibiting	HL	HL
Type HL + wear-inhibiting	HM	HLP
Type HL-P + detergent ("self-cleaning")	–	HLPD
Type HM + viscosity-improving	HV, HR	HVLP
Type HM + anti-stick-slip	HG	–

Figure 4 3 Types of Mineral oils

Water based hydraulic oils(Fire resistant hydraulic oils)- Water-based fluids are used for fire-resistance due to their high-water content. They are available as oil-in-water emulsions, water-in-oil (invert) emulsions and water glycol blends.

water-based fluids are used in applications when fire resistance is needed, these systems and the atmosphere around the systems can be hot. Elevated temperatures cause the water in the fluids to evaporate, which causes the viscosity to rise.

Occasionally, distilled water will have to be added to the system to correct the balance of the fluid. Whenever these fluids are used, several system components must be checked for compatibility, including pumps, filters, plumbing, fittings and seal materials. Water-based fluids can be more expensive than conventional petroleum-based fluids and have other disadvantages (for example, lower wear resistance) that must be weighed against the advantage of fire-resistance.

Description	ISO	DIN
Oil-in-water emulsions, mineral oil or synthetic ester	HFA E	HS-A
Water-based solutions of chemicals. Free of mineral oil. Water content > 80%	HFA S	-
Water-in-oil emulsions. Mineral oil content approx. 60%	HFB	HS-B
Water-polymer solutions. Water content > 35%	HFC	HS-C
Anhydrous synthetic fluids consisting of phosphates. Non-soluble in water.	HFD R	HS-D
Anhydrous synthetic fluids with a different origin, for example esters from carbolic acids	HFD U	HS-D

Figure 4 4 Water based hydraulic oils

Figure 4 4 gives the classification of water based oils as per ISO and Din standards. Synthetic fluids- are man-made lubricants and many offer excellent lubrication characteristics in high-pressure and high- temperature systems. Some of the advantages of synthetic fluids may include fire-resistance (phosphate esters), lower friction, natural detergency (organic esters and ester-enhanced synthesized hydrocarbon fluids) and thermal stability. The disadvantage to these types of fluids is that they are usually more expensive than conventional fluids, they may be slightly toxic and require special disposal, and they are often not compatible with standard seal materials. It is important to check the suitability of the oil considering the operating environments, types of seals chosen and other hydraulic elements of the system.

4.5 POINTS TO RECALL

1. Functions and characteristics of hydraulic oil.
2. The types of oil used in the system.

5. AN OVERVIEW OF A HYDRAULIC SYSTEM

5.1 LEARNING OUTCOMES

By the end of this chapter the student will understand

1. The general arrangement of a hydraulic system.
2. The common hydraulic elements and accessories used in a hydraulic system.

5.2 HYDRAULIC SYSTEM SET UP

In a hydraulic system, the hydraulic oil, we have discussed in the last chapter should move the hydraulic actuators. These hydraulic actuators can be either hydraulic cylinders (for linear reciprocatory motion) or hydraulic motors(For rotary motion).

These hydraulic actuators will be attached to the load that has to be pushed/pulled or pressed in the case of hydraulic cylinders and mixed or rotated if the actuator is a hydraulic motor.

The hydraulic oil has to be pumped to reach the hydraulic actuators.

Between the pump and the hydraulic actuators several other hydraulic components and accessories are used. Following table gives a brief classification of the elements used in a hydraulic system. This system is also referred as hydraulic power pack or hydraulic power unit.

Components/ Accessories	Examples	Functions
Components	Pumps	Pump hydraulic oil to flow to the actuators
	Valves	Different valves have different functions. Covered in the later part of this book.
	Actuators(Cylinders and motors)	Actuates the movements required
Accessories	Reservoir, filters, strainers, couplings, breather- fillers, bell housing, pressure gauge, oil level gauge etc.,	These accessories in aiding the hydraulic system in alignment, faltering, measuring functions. Covered in this book.
Pipe &Pipe Fittings	The piping work requires, Tees, bends, elbows etc.,	The fittings provide a leak free piping arrangement.

Table 5 1 Hydraulic system parts

Table 5 1 Hydraulic system parts Hydraulic system parts, gives an over view of parts used in a typical hydraulic system.

5.3 HYDRAULIC SYSTEM CIRCUIT LAYOUT

A schematic arrangement of a typical hydraulic system (power pack) is given in Figure 5 1

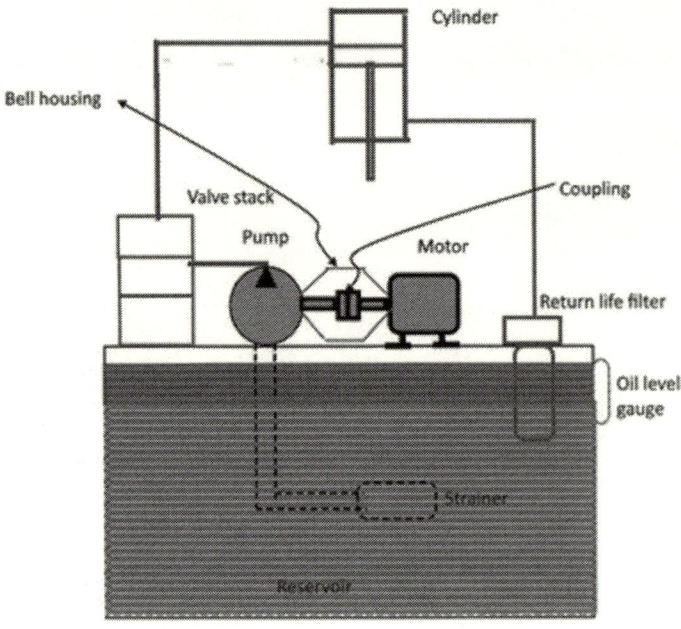

Figure 5 2 Hydraulic system arrangement

This is not a hydraulic circuit- but it is one step forward in understanding hydraulic circuits.

The main functions of the components are given below:

1. Pump/ Motor assembly – the pump is drawn with the hydraulic symbol. This pump is coupled to an electric motor by a coupling. To ensure that their shafts are in alignment a bellhousing is used.

2. Stack of valves- the pump outlet is connected by piping to a stack of different valves having different functions. The valves are mounted on a steel template (Referred as a manifold). This manifold is normally custom made depending on the circuit components.

3. The pump suction is inside the reservoir tank that contains the hydraulic oil. The piping is (shown in dotted lines) connected to the pump suction and to a strainer which is immersed in the hydraulic oil. The function of the strainer is that it does not allow iron fillings or nuts or bolts to get into suction side of the pump.(more of it later)

4. Double acting cylinder-The output of the valve stack is connected to the inlet port of the cylinder. The outlet of the cylinder is connected back to the tank through a return line filter. The return line filter removes the dust/dirt from the oil that has done the work in the cylinder. After filtration, the oil is back in the tank.

5. Breather filler-(Not shown in Figure 5 1) This is also mounted on top cover of the reservoir for refilling of oil and for the reservoir to breath – It is a process, when oil is taken up from the reservoir by the pump and after circulation through valves and cylinder, it comes back to refill the reservoir- this like breathing and air is required to fill and to be pushed out during the process. A breather filler takes care of this work.

6. Oil level gauge – indicates the oil level in the reservoir at any point in time.

7. Piping – From the suction strainer to the pump to stack of valves and to the cylinder and back to the tank, we would require pipes, fittings, hoses etc.,

8. Vent plug -generally located towards the bottom of the reservoir tank, a vent plug is given for easy drain of the oil.

5.4 POINTS TO RECALL

The general arrangement of hydraulic system and functions of each of the components used.

6. HYDRAULIC PUMPS

6.1 LEARNING OUTCOMES

By the end of the chapter, the student will understand

1. The function and types of pumping methods
2. Different types of pumps used in hydraulic systems.

6.2 TYPE OF PUMPS

We are all aware that pumps are used to transfer liquids from one location to another location. Nearly all pumps used, fall in the category of positive displacement pumps or non positive displacement pumps.

The most common type of pump used, especially in domestic application is centrifugal pumps – This falls in the non-positive displacement category.

There are many other versions of pumps – like turbine pumps, submersible pumps, jet pumps etc., These are chosen based on the application requirements. (All non positive displacement types)

The type of pump used in (oil) hydraulic application is referred as Positive displacement pumps. What is the significance of the name positive displacement pumps?

In this type of pumps (Positive displacement) the oil or liquid to be pumped gets transferred from the suction side to the delivery side – whatever be the pressure on the delivery side-the transferred volume of liquid or oil cannot slip bank to the suction side.

In a centrifugal pump, you can close the delivery side valve (if you have a shut off valve on the delivery side of the pump) and the volume of liquid taken from the suction side for transfer, will be churning back to the suction side. If you keep the centrifugal pump running, then, the pump keep churning the liquid inside and there will not be any major damage to the centrifugal pump.

In a positive displacement pump, if you close the delivery side, when the pump is transferring the liquid, as the transferred liquid, cannot go back to suction side, the pressure will build up and there is a possibility of the piping getting burst or the pump will be damaged. The delivery per cycle remains almost constant, regardless of changes in pressure against which the pump is working.

Please refer Figure 6 1.The centrifugal has varying flow depending on pressure or head, whereas the Positive displacement pump has more or less constant flow regardless of pressure. (pd Vs centri.pdf, 2007)

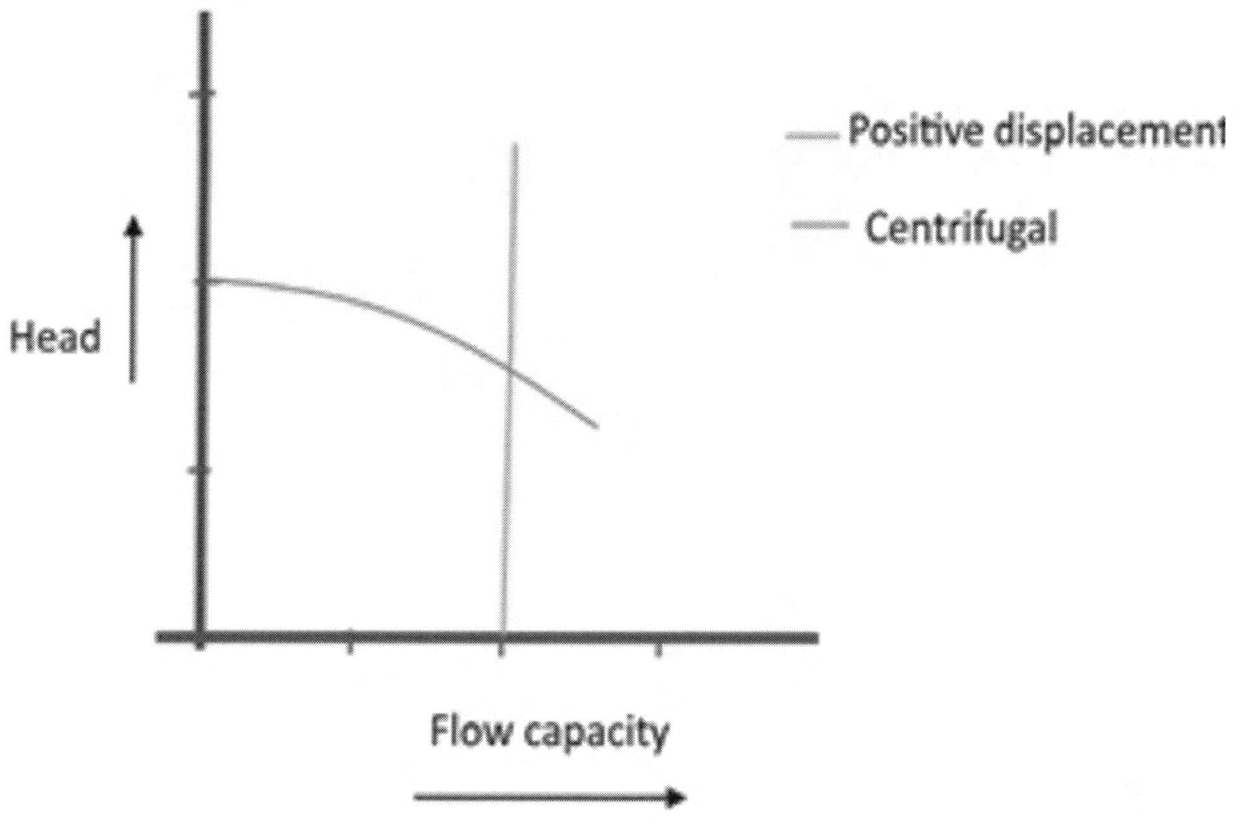

Figure 6 1 Centrifugal Vs Positive displacement pump performance

In most hydraulic systems, we need Positive displacement pumps so that the liquid transfers energy and make the actuator to work against the load.

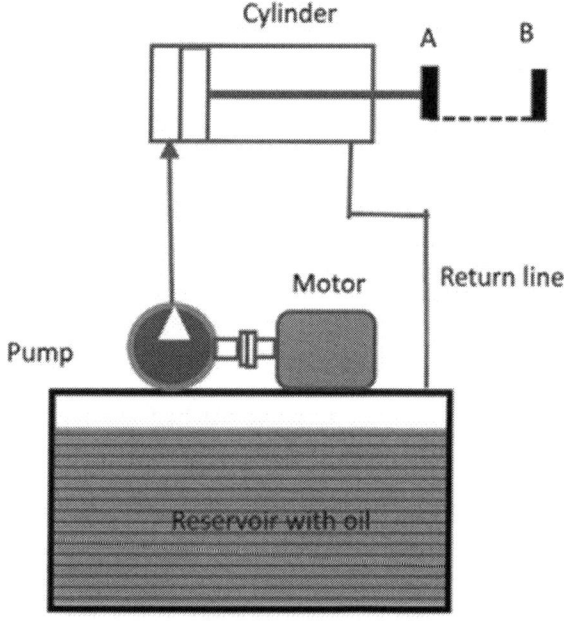

Figure 6 2 Positive displacement pump for hydraulic powerpack

Please refer Figure 6 2.A simple hydraulic system arrangement with essential components are shown.

In the above scheme shown, the load attached to the cylinder shaft has to be shifted from A to B. If we do not use positive displacement pump, then, it is possible that the oil from the pump is not transferring the full volume of oil intended (as per the calculated capacity). This would mean that the load pressure is making the centrifugal pump to slip some volume back into suction side and the load movement will not be smooth or effective.

6.3 PUMP PARAMETERS

The following parameters are essential in pump selection.

Flow rate -is the rate of output of the pump. This is given in cubic centimeters(cc) per minute(Cubic feet per minute) or Liters per minute- l/min(gallons per minute-gpm).

Manufacturers at times specify the pump flow rate as cc / revolution. If for example, the flow rate given is 4cc/revolution for a particular pump, to get l/min, please multiply this flow rate by the speed of the prime mover(Say- motor).If the motor speed is 1500 revolutions per minute(rpm), then the capacity of the pump is 4cc X 15000rpm= 6000cc.That is 6l/min(1 l/min=1000cc).

However the pump may not be exactly delivering6l/min – The reason being that the efficiency of the pump may be only 70% and in that case, it would be delivering 6X0.7 = 4.2l/min.

Volumetric efficiency –. The actual flow rate divided by the theoretical flow rate is the 'volumetric efficiency'. Most positive displacement pumps have a volumetric efficiency between 85 and 98 percent.

For example, a pump may be nominally rated as a 50 gpm (13.6 m3/h) unit. Under no load conditions, the pump can deliver considerably more than 50 gpm, but less than 50 gpm at its rated operating pressure. As mentioned above due to the efficiency of the pump being less, we get, reduced output.

Volumetric efficiency = Actual output / Theoretical output.

Most of the pump manufacturers give the volumetric efficiency curves for the type of pumps manufactured and for different rpms.

Power required by the pump for driving (Prime mover power)- It should be noted that the efficiency itself can vary with pressure. The designer of hydraulic system should look at the characteristic curves of the pump (From the manufacturers data catalogue for the pumps) and find out the flow rate at a particular pressure and the power absorbed by the pump.

There is also a formula for the power absorbed by the pump.

Power absorbed by the pump in KW =(P Q)/600

Where P = Maximum pressure of the system in Bar or psi

Q = lpm of the pump chosen.

Let us look at the following example:

A hydraulic pump delivers 12 L of fluid per minute against a pressure of 200 Bar. (a) Calculate the hydraulic power. (b) If the overall pump efficiency is 60%, what size of electric motor would be needed to drive the pump?

Solution:

Hydraulic power(Fluid Horse power- FHP) is given by PQ/600 : P=200 Bar;

Q= 12l/min: 200 X12 /600 =4 kW

(b) We have Hydraulic Power Electric motor power (power input)

= Hydraulic power/Overall efficiency

= 4/0.6 = 6.6 KW. we must choose the nearest next higher size motor KW acvailable in the market – which is 8 KW.

From the above it is clear that the overall efficiency of a positive displacement pumps is:

often called "mechanical efficiency" in PD pumps, is a ratio of useful hydraulic power (FHP) transmitted to the fluid exiting the pump divided by total power (BHP) absorbed by the pump.

6.4 PUMP CLASSIFICATION

Positive displacement pumps are of two categories.

1. Reciprocating pumps - includes plunger, diaphragm, piston, hand operated and many others -The principle is to use a repetitive reciprocating mechanism to expand and contract the chamber at regular intervals. Reciprocating pumps incorporate one or more sets of check valves at the inlet and outlet of the pump to help guild the liquid through the pump and to prevent reverse flow.

2. Rotary pumps. - Includes gear, screw, vane, lobe, and piston types — The principle here is to use use rotating parts to move the liquid in and out of the pump chamber. Some rotary pumps, such as gear pumps, must have very tight clearance between the rotating elements and the walls of the chamber, and between the rotating parts, which means they generally can't be used to pump large Solids or abrasive fluids that may wear the parts. Other types such as lobe pumps, screw pumps are designed to move liquids containing Solids.

Another way of classification is as given below:

Based on construction:

1. Gear pumps
2. vane pumps
3. Axial piston pumps
4. radial piston pumps

Based on delivery of oil flow

1. Fixed displacement pumps
2. Variable displacement pumps.

If we have to discuss all types of pumps in this book, then it will be a book on pumps. As our focus is more on learning about hydraulic systems and circuits, we shall discuss the most popular types of pumps. Most of the manufacturers would willingly give all details of any particular pump manufactured by them. Hence our discussions would cover the essential constructional and operational features of the pumps.

6.5 RECIPROCATING PUMP – OPERATING PRINCIPLE

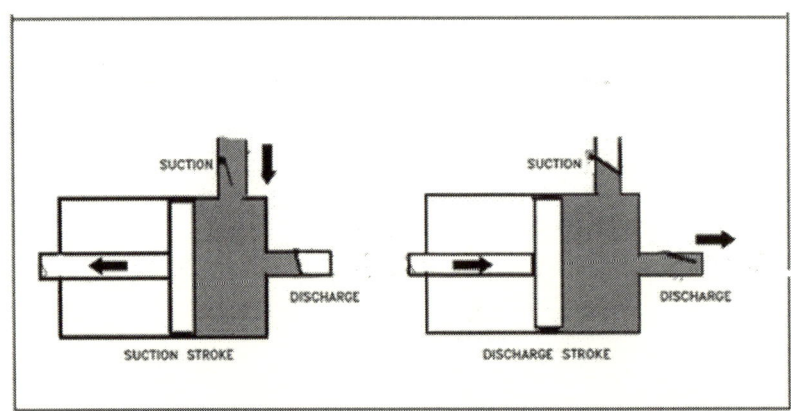

Figure 6 3 Reciprocating pump- operating Principle

All positive displacement pumps operate on the same basic principle. This principle can be most easily demonstrated by considering a reciprocating positive displacement pump consisting of a single reciprocating piston in a cylinder with a single suction port and a single discharge port as shown in Figure 6 2. Check valves in the suction and discharge ports allow flow in only one direction.

Please refer Figure 6 2. During the suction stroke, the piston moves to the left, causing the check valve in the suction line between the reservoir and the pump cylinder to open and admit the oil r from the reservoir. During the discharge stroke, the piston moves to the right, seating the check valve in the suction line and opening the check valve in the discharge line. The volume of oil moved by the pump in one cycle (one suction stroke and one discharge stroke) is equal to the change in the liquid volume of the cylinder as the piston moves from its farthest left position to its farthest right position.

6.6 ROTARY PUMPS

6.6.1 External gear pumps

Gear pumps can be further classified based on design as external gear pumps and internal gear pumps.

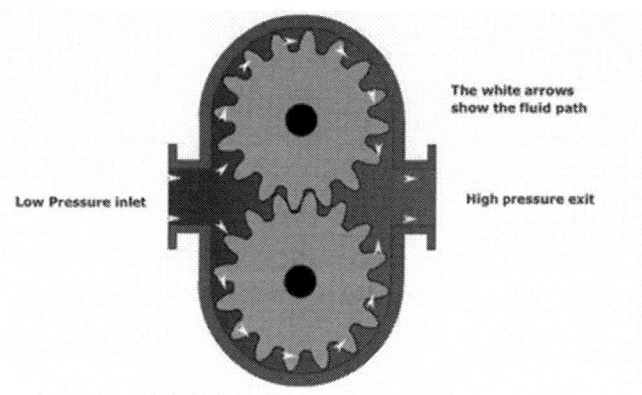

Figure 6 4 External gear pump

Please refer Figure 6 4. consists of two spur gears meshing together and revolving in opposite directions within a casing. Only a fraction of a millimeter(thousandths of an inch clearance exists between the case and the gear teeth extremities.

Any liquid that fills the space bounded by two successive gear teeth and the case must follow along with the teeth as they revolve. When the gear teeth mesh with the teeth of the other gear, the space between the teeth is reduced, and the entrapped liquid is forced out the pump discharge pipe. As the gears revolve and the teeth disengage, the space again opens on the suction side of the pump, trapping new quantities of liquid and carrying it around the pump case to the discharge. As liquid is carried away from the suction side, a lower pressure is created, which draws liquid in through the suction line.

Figure 6 5 Gear pump cross section

With the large number of teeth usually employed on the gears, the discharge is relatively smooth and continuous, with small quantities of liquid being delivered to the discharge line in rapid succession. If designed with fewer teeth, the space between the teeth is greater and the capacity increases for a given speed; however, the tendency toward a pulsating discharge increases. In all simple gear pumps, power is applied to the shaft of one of the gears, which transmits power to the driven gear through their meshing teeth.

There are no valves in the gear pump to cause friction losses as in the reciprocating pump. The high impeller velocities, with resultant friction losses, are not required as in the centrifugal pump. Therefore, the gear pump is well suited for handling viscous fluids such as fuel and lubricating oils. External gear pumps are used in industrial and mobile (e.g. log splitters, lifts) hydraulic applications. Typical applications are lubrication pumps in machine tools, fluid power transfer units and oil pumps in engines. These pumps are normally used in applications with maximum pressure upto 200 Bar.

6.6.2 Lobe pump

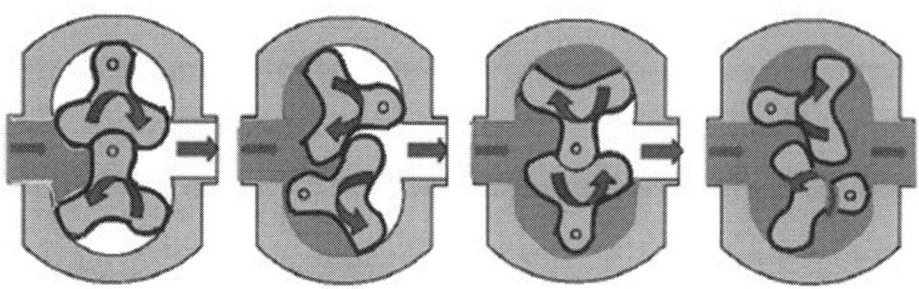

Figure 6 6 Lobe pump

Lobe pumps are similar to external gear pumps in operation in that fluid flows around the interior of the casing. As the lobes come out of mesh, they create expanding volume on the inlet side of the pump. Liquid flows into the cavity and is trapped by the lobes as they rotate.

Liquid travels around the interior of the casing in the pockets between the lobes and the casing -- it does not pass between the lobes. Finally, the meshing of the lobes forces liquid through the outlet port under pressure.

Figure 6 7 lobe pump construction

Lobe pumps are frequently used in food applications because they handle Solids without damaging the product. Particle size pumped can be much larger in lobe pumps than in other PD types. Unlike external gear pumps, the lobes do not make contact, and clearances are not as close as in other PD pumps, this design handles low viscosity liquids with diminished performance. Normally the pressure range is less- less than 100 Bar.

6.6.3 Internal gear pumps

The internal gear pumps are often used on thin liquids such as Solvents and fuel oil. they find application in pumping thick liquids such as asphalt, chocolate, and adhesives.

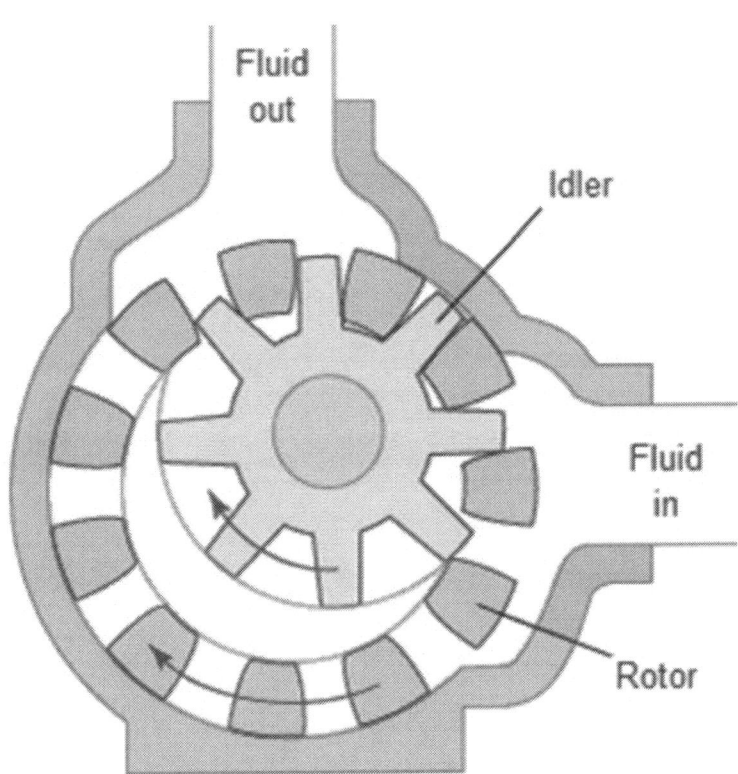

Figure 6 8 Internal gear pump.

They are also used in non-mobile hydraulics (e.g. machines for plastics and machine tools, presses, etc.) and in vehicles that operate in an enclosed space (electric fork-lifts, etc.).

The principal of operation is as follows.

Liquid enters the suction port between the rotor (large exterior gear) and idler (small interior gear) teeth. The arrows indicate the direction of the pump and liquid.

Liquid travels through the pump between the teeth of the "gear-within-a-gear" principle. The crescent shape divides the liquid and acts as a seal between the suction and discharge ports.

The pump head is now nearly flooded, just prior to forcing the liquid out of the discharge port. Intermeshing gears of the idler and rotor form locked pockets for the liquid which assures volume control.

Generally, the internal gear pumps are available for maximum pressure: up to 300 Bar (dependent on nominal size)

6.6.4 Vane pump

A rotary vane pump is a positive-displacement pump that consists of vanes mounted to a rotor that rotates inside of a cavity.

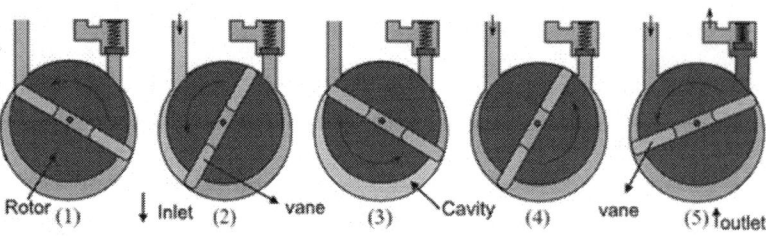

Figure 6 9 Vane pump working principle

Please refer Figure 6 9 shows a rotor (Red color). The arrow shows the direction of rotation. The rotor has two slots for vanes (Pink color). The vanes slide into these two slots. These vanes are spring loaded. This means the vanes would be pushed out if there is room for it to slide out in to the cavity.

In (1),the vane blocks entry of the fluid. The rotor starts rotating with the vane. As there is space for the vane to slide out, during the rotational movement of the rotor, it extends to touch the inner side of the cavity (referred as cam ring inner surface).

As the rotor rotates, there is low pressure created just above the vane(marked (2)) and the fluid(oil) enters the cavity.

In (3), the second vane pushes this oil through the spring-loaded valve to delivery side. As the rotation is continuous, there is a smooth transferring of oil from inlet(suction) to outlet(Delivery).If you observe Figure 6 10, the housing may be eccentric with the center of the rotor, or its shape may be oval.

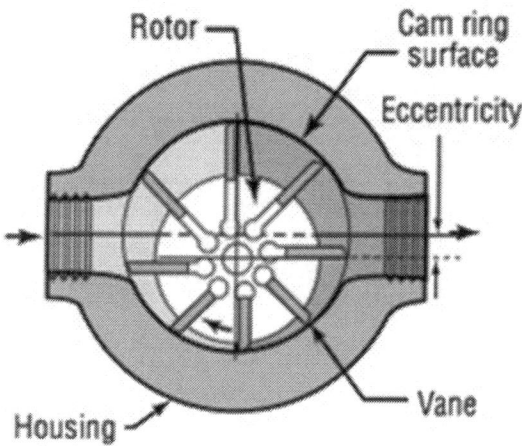

Figure 6 10 vanes Rotor and eccentricity (Note suction/delivery locations)

In some designs, centrifugal force holds the vanes in contact with the housing(instead of springs), while the vanes are forced in and out of the slots by the eccentricity of the housing.

Figure 6 11 Vane in the rotor slots and cavity eccentricity

Vane pumps are noted for their dry priming, ease of maintenance, and good suction characteristics over the life of the pump.

The pump illustrated in Figure 6 10 is unbalanced, because all of the pumping action occurs in the chambers on one side of the rotor and shaft. This design imposes a side load on the rotor and drive shaft. This type vane pump has a circular inner casing.

Some vane pumps provide a balanced construction in which an elliptical casing forms two separate pumping areas on opposite sides of the rotor(instead of a circular one), so that the side loads cancel out. Please refer Figure 6 12. This is Balanced vane pump.

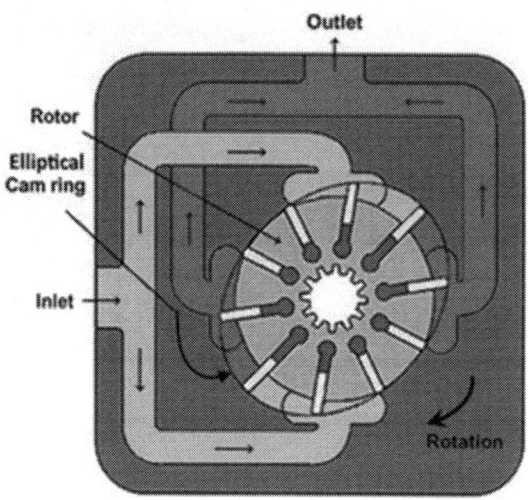

Figure 6 12 Balanced vane pump with elliptical cam ring.

Balanced vane pumps come only in fixed displacement designs. This means if the pump is designed for 6l/min at a particular speed, it would only give that volume of oil (Fluid) at the speed. However, in certain pumps it is possible to change the displacement of oil(Fluid) by attaching an additional feature referred as Pressure compensator. This is possible only in unbalanced vane type of pumps only.

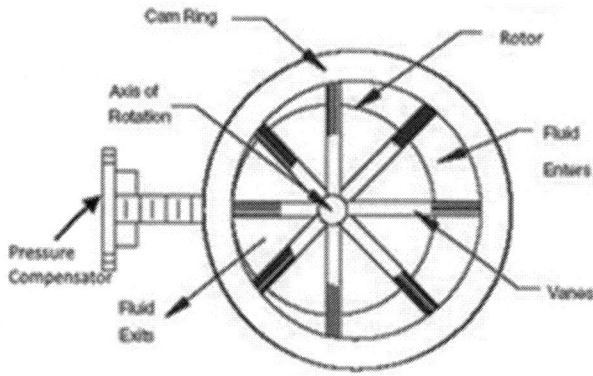

Figure 6 13 variable vane pump with pressure compensation

In a variable-volume unbalanced design, the displacement can be changed through an external control such as a handwheel or a pressure compensator.
The pressure compensator control moves the cam ring to change the eccentricity between the ring and rotor, thereby changing the size of the pumping chamber and thus varying the displacement per revolution. When pressure is high enough to overcome the compensator spring force, the cam ring shifts to decrease the eccentricity. Adjustment of the compensator spring determines the pressure at which the ring shifts. In Figure 6 13, the inlet and outlets are not shown. The image is to make the

students understand the pressure compensation and variable delivery of vane pumps.

The operating pressure of vane pumps does not normally exceed 175 Bar. However, in specially designed vane pumps the operating pressure may go over 200 Bar and up to 300 Bar.

6.7 PISTON PUMPS

There are three types of Piston pumps-

1. Radial Piston pumps
2. Axial Piston Pumps
3. Bent axis pumps.

6.7.1 Radial Piston pumps

Hydraulic piston pumps can handle large flows at high hydraulic system pressures. Typical applications are mobile and construction equipment, marine auxiliary power, metal forming and stamping, machine tools and oil field equipment, Test rigs, automotive sector (e.g., automatic transmission, hydraulic suspension control in upper-class cars), plastic- and powder injection molding.

In these pumps, the pistons accurately slide back and forth inside the cylinders that are part of the hydraulic pump.

Please refer Figure 6 14. In this pump design, a pentagon shaped cam is having five blocks(Black color) attached to its five surfaces. This cam rotates eccentrically in the circular (yellow) cavity.

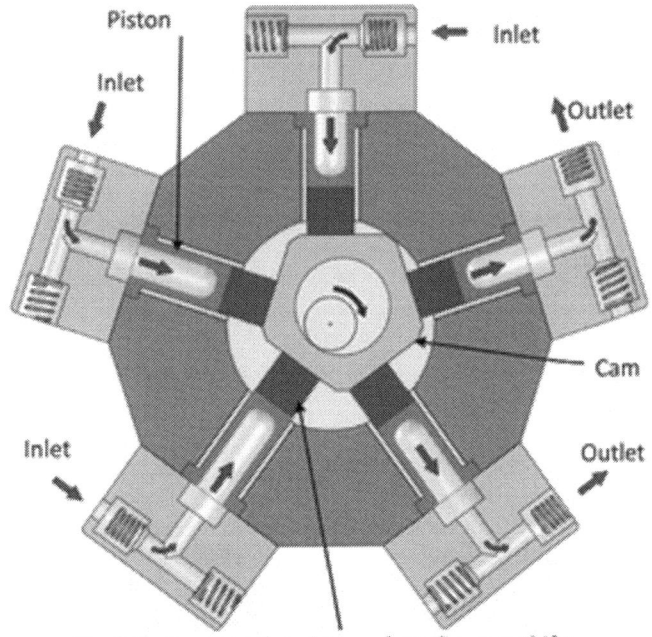

Figure 6 14 Radial Piston pumps

This eccentricity determines the stroke of the pumping piston. If the cam is away from the piston, it draws the piston inwards, causing a suction of oil(fluid). As the cam rotates another surface of cam surface pushes the piston away, causing delivery of the pump.

Each piston has suction and delivery ports and each pump. Externally, It is possible to connect all suctions together and delivery sides together. These kinds of piston pumps have advantages as listed:

high efficiency, high pressure (up to 1,000 Bar),low flow and pressure ripple (due to the small dead volume in the workspace of the pumping piston),low noise level and high reliability.

6.7.2 Axial Piston pumps

As the name suggests here the pistons are in line with the axis of the driving prime mover(Motor or an Engine).

This pump has a number of pistons arranged in a circular array within a housing which is commonly referred to as a cylinder block, rotor or Barrel. This cylinder block is driven to rotate about its axis of symmetry by an integral shaft that is, more or less, aligned with the pumping pistons/ Please refer the exploded view given below.

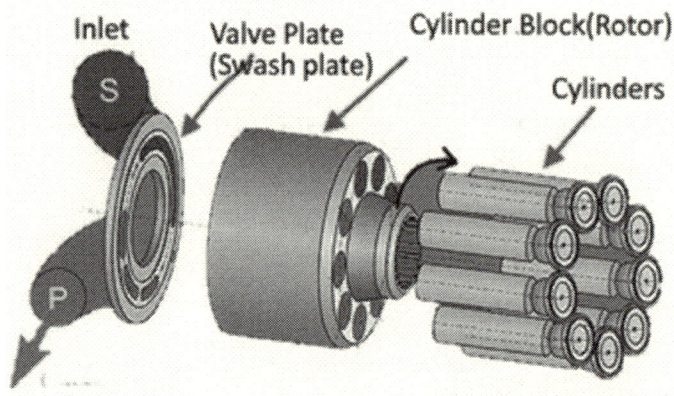

Figure 6 15 Exploded view of Axial piston pump

In Figure 6 16 the assembled axial piston pump is shown. Please note that in the assembled image the pistons are shown inside the cylinder block cavities and for the sake of simplicity only two cylinders are shown.

Further, the swash plate is at an angle (shown vertical in Figure 6 15 Exploded view of Axial piston pump) . The suction and delivery ends are shown as an extension of cylinder ports. Normally they are clustered together (incase of number of cylinders- all suctions together and all deliveries together and taken through the red and blue sgaded portions of the valve plate.

As the cylinder block rotates, the exposed ends of the pistons are constrained to follow the surface of the cam plane. Since the cam plane(swash plate) is at an angle to the axis of rotation, the pistons must recciprocate axially as they press about the cylinder block axis. The axial motion of the pistons is sinusoidal. During the rising portion of the piston's reciprocation cycle, the piston moves toward the valve plate. Also, during this time, the fluid trapped between the buried end of the piston and the valve plate is vented to the pump's discharge port through one of the valve plate's semi-circular ports - the discharge port. As the piston moves toward the valve plate, fluid is pushed or displaced through the discharge port of the valve plate.

When the piston is at the top of the reciprocation cycle (commonly referred to as top-dead-center or just TDC), the connection between the trapped fluid chamber and the pump's discharge port is closed. Shortly thereafter, that same chamber becomes open to the pump's inlet port. As the piston continues to press about the cylinder block axis, it moves away from the valve plate thereby increasing the volume of the trapped chamber. As this occurs, fluid enters the chamber from the pump's inlet to fill the void. This process continues until the piston reaches the bottom of the reciprocation cylinder - commonly referred to as Bottom-Dead-Center or BDC. At BDC, the connection between the pumping chamber and inlet port is closed. Shortly thereafter, the chamber becomes open to the discharge port again and the pumping cycle starts over.

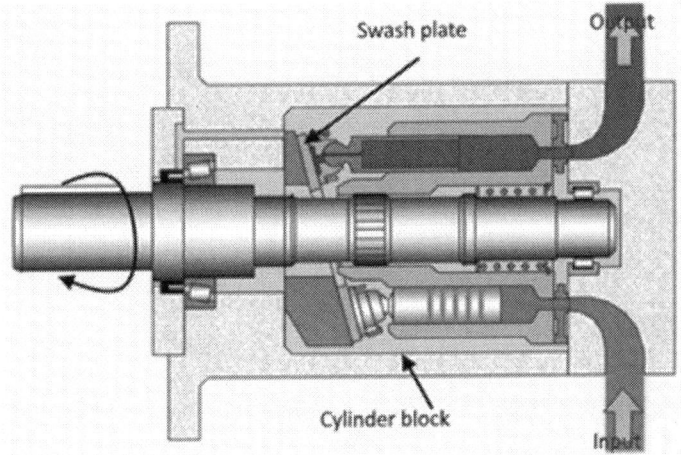

Figure 6 16 Axial piston pump with swash plate at an angle

if cam plane (swash plate) is set parallel to the axis of rotation, there is no movement of the pistons in their cylinders. Thus there is no output. Movement of the swash plate controls pump output from zero to maximum.

In a typical pressure-compensated pump, the swash plate angle is adjusted through the action of a valve which uses pressure feedback so that the instantaneous pump output flow is exactly enough to maintain a designated pressure. If the load flow increases, pressure will momentarily decrease but the pressure-compensation valve will sense the decrease and then increase the swash plate angle to increase pump output flow so that the desired pressure is restored. In reality most systems use pressure as a control for this type of pump. The operating pressure reaches, say, 200 Bar (20 MPa or 2900 psi) and the swash plate is driven towards zero angle (piston stroke nearly zero) and with the inherent leaks in the system allows the pump to stabilize at the delivery volume that maintains the set pressure.

As demand increases the swash plate is moved to a greater angle, piston stroke increases and the volume of fluid increases; if the demand slackens the pressure will rise, and the pumped volume diminishes as the pressure rises. At maximum system pressure the output is once again almost zero. If the fluid demand increases beyond the capacity of the pump to deliver, the system pressure will drop to near zero. The swash plate angle will remain at the maximum allowed, and the pistons will operate at full stroke. This continues until system flow-demand eases and the pump's capacity is greater than demand. As the pressure rises the swash-plate angle modulates to try to not exceed the maximum pressure while meeting the flow demand

Despite the problems indicated above this type of pump can contain most of the necessary circuit controls integrally (the swash-plate angle control) to regulate flow and pressure, be very reliable and allow the rest of the hydraulic system to be very simple and inexpensive.

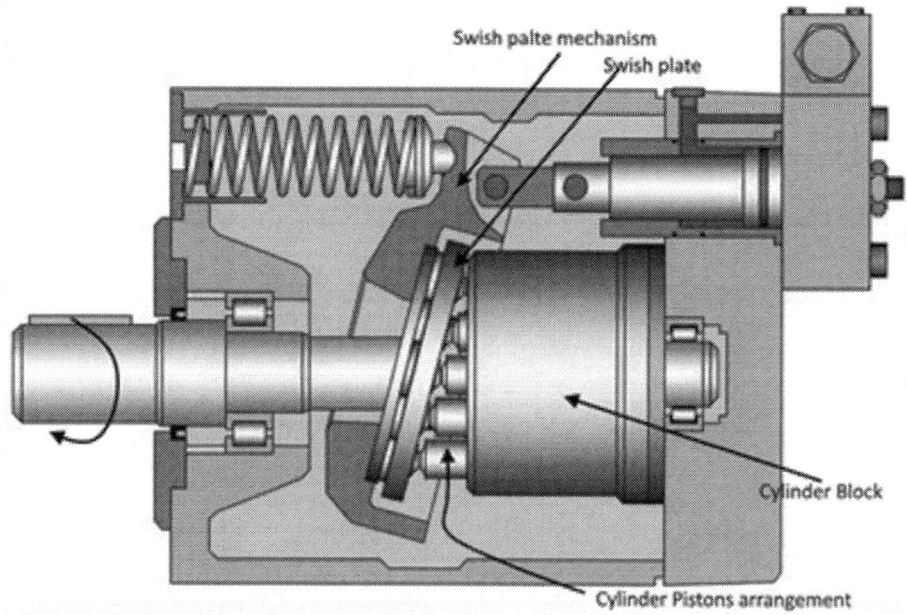

Figure 6 17 Axial piston pump construction

These pumps are available for pressures up to 300Bar. Axial piston pumps are used to power the hydraulic systems of jet aircrafts . earth moving plants etc.,

6.7.3 Bent axis pump

In this construction of axial piston pumps, the reciprocating action of the pistons is obtained by bending the axis of the cylinder block. The cylinder block rotates at an angle which is inclined to the drive shaft. The cylinder block is turned by the drive shaft through a universal link.

The cylinder block is set at an offset angle with the drive shaft. The cylinder block contains a number of pistons along its periphery. These piston rods are connected with the drive shaft flange by ball-and-socket joints. These pistons are forced in and out of their bores as the distance between the drive shaft flange and the cylinder block changes.

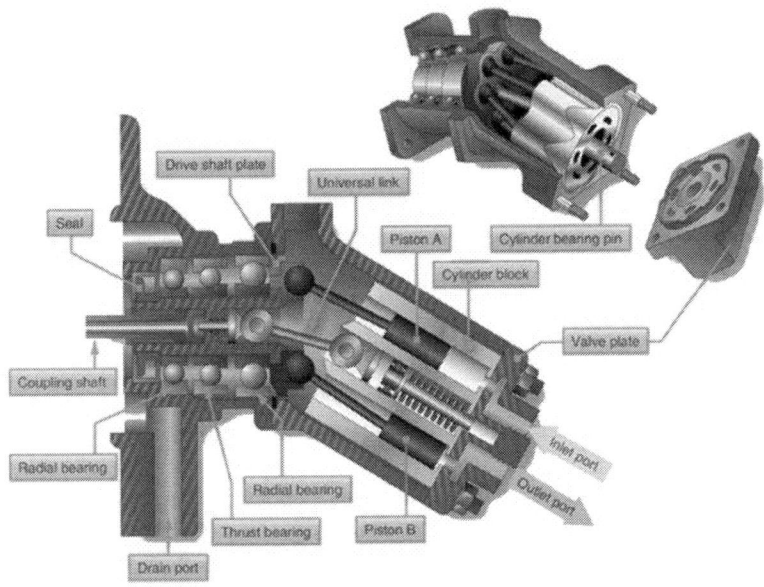

Figure 6 18 Bent axis pump construction

A universal link connects the block to the drive shaft, to provide alignment and a positive drive. The fixed displacement units are usually provided with 23° or 30° offset angles while the variable displacement units are provided with a yoke and an external control mechanism to change the offset angle.

These pumps are available for maximum pressure up to 350- 400 Bars. These pumps find applications in mobile hydraulics and they are used to in mobile and construction equipment, winches, ship-cranes and all kinds of heavy-duty hydraulic equipment for offshore and onshore operations.

6.8 USAGE OF PUMPS IN OPEN AND CLOSE HYDRAULIC SYSTEMS

The axial pumps predominantly find application in closed loop hydraulic circuits.

6.8.1 Closed Loop Hydraulic system

In case of closed loop hydraulic system, Hydraulic fluid will flow from pump to actuators. Hydraulic fluid will enter at inlet of pump after passing through the actuators. One hydraulic pump might be used for driving the multiple hydraulic motor in case of closed hydraulic system. Hydraulic pump will deliver the fluid to actuators and will receive the same quantity of fluid from its inlet port from the actuators for smooth operation of the system. However, one feed pump is always provided with closed loop hydraulic system in order to make-up the fluid in closed loop circuit. Feed pump will be a fixed displacement having capacity approximate 15 % of main pump.

6.8.2 Open loop Hydraulic system

n case of open loop hydraulic system, hydraulic pump will suck fluid from reservoir and will deliver the high pressure fluid to the actuators i.e. hydraulic cylinders via passing through pressure relief valve , pressure regulating and flow control valve. After passing from actuators, fluid will go to heat exchanger and then reservoir and again pump will take fresh fluid from reservoir via its inlet port and this process will be continuous during operation of hydraulic actuators.

These concepts will be shown as hydraulic circuits in later chapters of this book. In fact in circuit building chapter, we will start the discussions with open loop circuits.

The reference of open loop and close circuits arose here because axial piston pumps are used in many cases of close loop circuits where the pump feeds a hydraulic motor and the output from the motor goes to the pump.

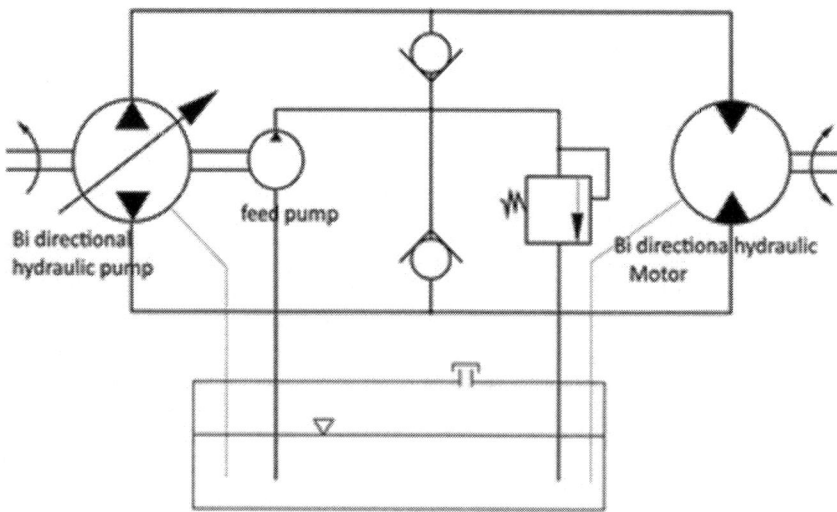

Figure 6 19 Example of close loop hydraulic circuit.

6.9 POINTS TO RECALL

1. The students should remember the following symbols:

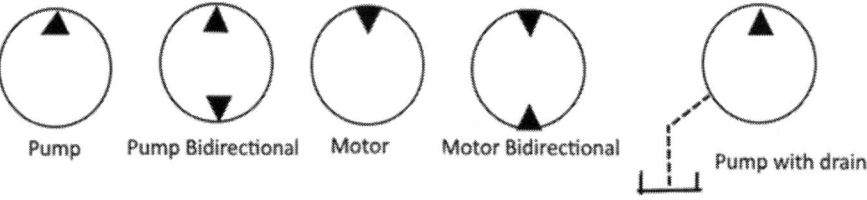

Figure 6 20 Pump symbols for hydraulic circuits

2. It is important for the students in principle to remember the general constructional features, rather than the details- and the pressure range and the application for which the types are mostly used.

7. HYDRAULIC ACTUATORS

7.1 LEARNING OUTCOMES:

By the end of this chapter, the students will understand

1. The types of hydraulic cylinders /classification.

2. Essential specifications/ Parameters of hydraulic cylinders.

7.2 CLASSIFICATION OF HYDRAULIC ACTUATORS

Hydraulic actuators are classified as follows.

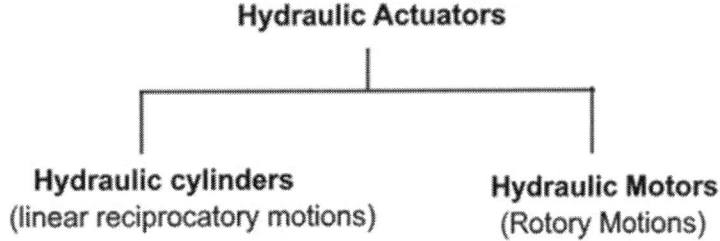

Figure 7 1 Classification of Hydraulic Actuators

7.2.1 Hydraulic cylinders

Types of hydraulic cylinders – There are different types of hydraulic cylinders in use for different applications and they are categorized below in Figure 7 2 Types of cylinders.

The so many types may look daunting for the first time students- But, the most popular are the double acting cylinders of full bore type. In the machine tool applications, where the operating pressure is 100 Bar or less,

A tie rod type of cylinder construction is used.

For higher pressures, welded type / bolted construction hydraulic cylinders are mostly used. The images of each type of cylinders are given after the classification details.

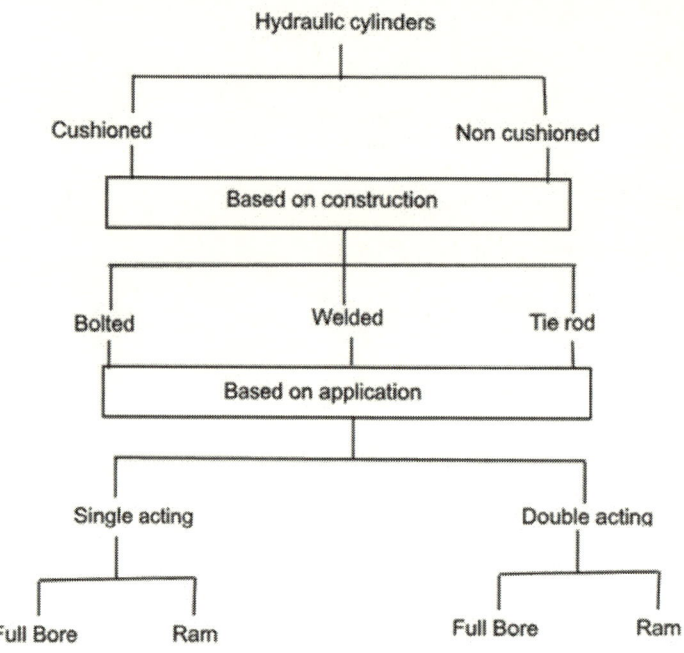

Figure 7 2 Types of cylinders

7.2.2 Working Principle of Hydraulic cylinders

Please refer Figure 7 3. The picture shows a double acting cylinder.

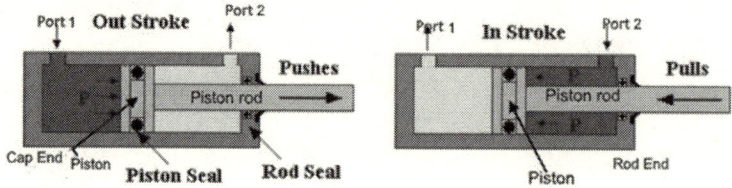

Figure 7 3 Hydraulic cylinder working principle

The essential parts are marked in the cylinder. Initially the oil enters through the port1 in left side figure (marked red) and pushes the piston with pressure P. As the piston starts moving/extending (along with piston rod connected to it- From Cap end), the oil on the other side of the piston (yellow) gets pushed out of the port 2. Generally, if there is a load attached to the piston rod, it gets pushed by the movement of the piston and the piston rod.

Once the piston reaches the other end(Rod end), by means of a special valve (Direction control valve) the oil flow enters Port 2 and start pushing the piston from piston rod side. Now the oil this rod side is shown in red. As the piston rod retracts the oil that had earlier entered (now shown in yellow) gets pushed out through port1.

Initial oil entry Through Port 1- The oil pushes the full piston area (referred as full bore area). By a direction flow control valve, we make oil enter port 2 to put pressure on the rod side of the piston. This area is referred as annulus area.(Annulus area= Full bore area- rod area)

[All the terms mentioned in bold letters will be frequently referred in this book]

If P = Pressure on the full bore of the cylinder.

And A = Piston full bore area

And a= Rod area, Then,

Force exerted by the oil on full bore area = P * A

Force exerted by oil on annulus area = P(A-a)

The speed of movement of the piston/ piston rod depends on the flow rate of oil entering the cylinder.

Q = Flow rate on full bore side in cubic cm/ sec = A(cm^2) * Velocity(cm/sec)

Q= Flow rate on annulus area side in cubic cm/sec=(A-a)cm^2 *Velocity(cm/sec)

7.2.3 Specifications of Hydraulic cylinders

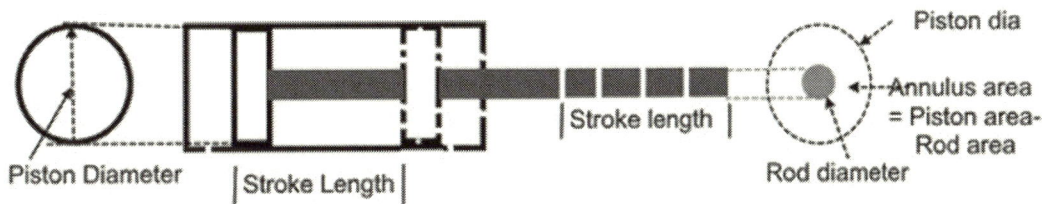

Figure 7 4 Specifications of a hydraulic cylinder

Figure 7 4 shows the stroke length, piston diameter and rod dimeters are shown.

The stroke length is the length of travel of the piston. It is the same length by which the piston rod extends from its initial position. As the piston extends, the oil extends pressure on full face of the piston or the full bore area. When, the piston is retracting, the oil exerts pressure on the annulus area.

7.2.4 Symbols of Hydraulic cylinders for hydraulic circuit drawing

Please refer symbols of hydraulic cylinders that are frequently used in hydraulic cylinders.

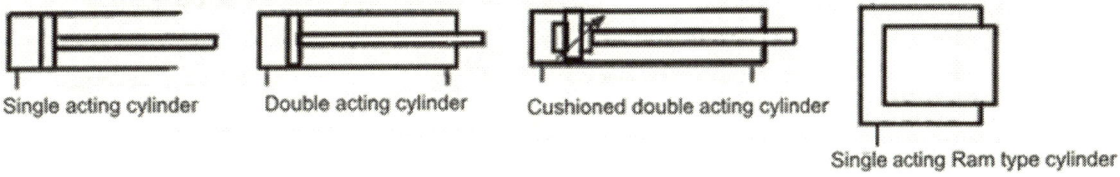

Figure 7 5 Hydraulic cylinder types and symbols.

7.2.5 Cushioned and non-cushioned cylinders

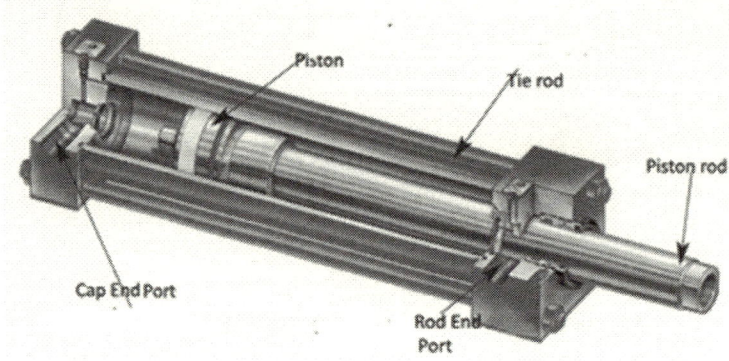

Figure 7 6 Tie rod cylinder

Please refer Figure 7 6

It shows a tie rod type of a cylinder and also shows the cap end and tie rod ends of the cylinder. The piston rod has to move between the two ends of the cylinder mentioned. It would be advantageous if the cylinder reduces its speed of movement whenever it approaches the ends of its travel.

Reducing the piston velocity as it approaches the cap end lowers the stresses on cylinder components and reduces vibration transmitted to the machine structure, if at all cylinder impacts the cylinder end. It is possible to reduce the speed of the piston movement by 'Cushioning' the cylinder.

This is done by narrowing the port size and providing a valve that would restrict the oil flow getting out of the port. If the cylinder cushioning (Restricting the outflow of oil from the cylinder) is adjustable, then the speed of approach of the piston also becomes adjustable. Please refer the symbol of cushioned cylinder and note the arrow over the piston. This means it is a cylinder with adjustable cushioning.

Please refer Figure 7 7.

Cylinders are manufactured in three most popular ways- Tie rod type- where the cap end and rod ends are tied together by tie rods(shown in extreme left(- The bolted construction(Middle figure) – In this type the cylinder tubes are manufactured with flanged ends and bolts are then used to fit the cap end and rod ends. In the welded construction welding is resorted for joining the ends.

Tie rod cylinders are used for low pressure applications up to 70 Bar pressure. For higher pressures than 70 Bar, normally a bolted or welded construction is selected.

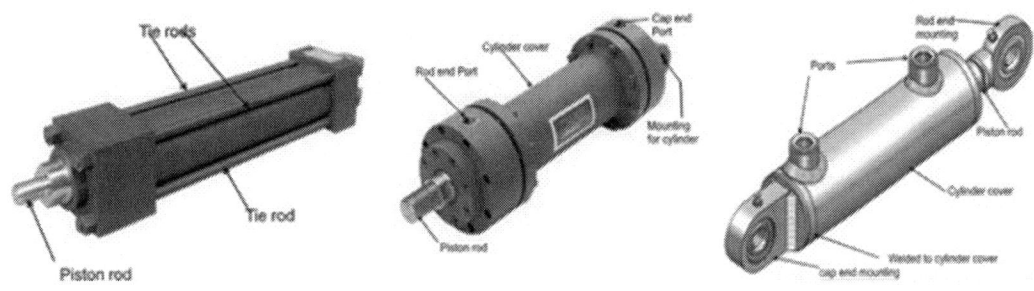

Figure 7 7 Tie rod, Bolted and welded type cylinders

Double acting and Single acting hydraulic cylinders

A double-acting hydraulic cylinder has a port at each end, supplied with hydraulic fluid for both the retraction and extension of the piston. This is shown in Figure 7 3 . The hydraulic fluid (Oil) is required for both the movements.

In a single acting cylinder, it is not so.

In a single acting cylinder, only for one direction of motion oil is used as shown in Figure 7 8

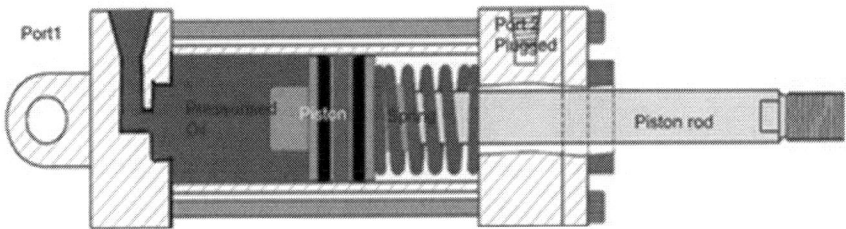

Figure 7 8 Single acting cylinders

In the figure shown above, the spring is used in the annulus area(rod side).It means that, when Port 1 is free and open to the reservoir, then, the spring would push the piston and the piston rod back into cap end position.

In case the spring is used in the full bore side(Piston face) Then the piston will be in rod end position in the beginning and we shall pump oil through port 2 and move the piston/piston rod in to retracted position. That is it will move to the cap end position against the spring force.

At times, in single acting, we might use the load weight itself to retract the piston, instead of a spring as shown in Figure 7 8

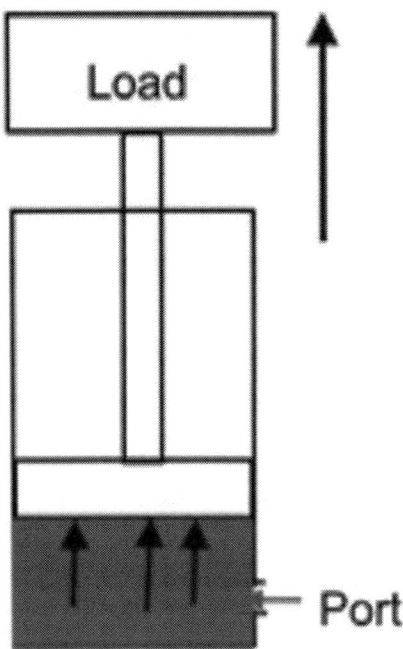

Figure 7 9 Single acting cylinder

At the time of retraction, there should not be any flow of oil through the port and it should be open to the reservoir (which can be done with direction control valve).

7.2.6 Ram Cylinder

A ram-type cylinder is a cylinder in which a cross-sectional area of a piston rod is more than one-half a cross-sectional area of a piston head. In many cylinders of this type, the rod and piston heads have equal areas. A ram-type actuating cylinder is used mainly for push functions rather than pull.

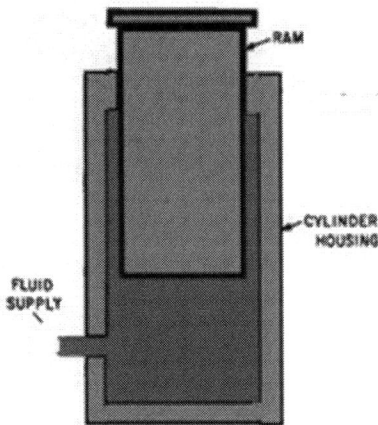

Figure 7 10 ram type - Single acting

Please refer Figure 7 10. This is a single acting Ram type cylinder. Please note the size of ram dia. The cylinder has only one port for admission of oil in to the cylinder. Even in a double acting ram type, the cylinder will have two ports, but the ram is considerably more than the regular piston rod proportion.

7.2.7 Telescopic cylinder

This has not been indicated in the regular types of cylinders mentioned in Figure 7 2.this type of cylinder is more popular in mobile hydraulics.

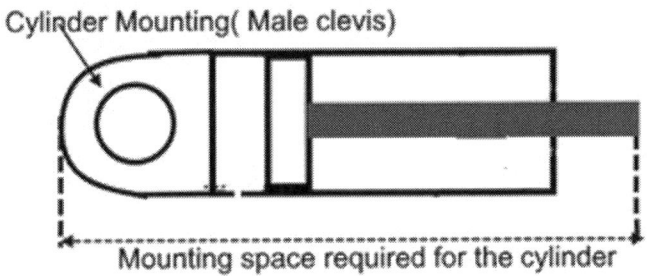

Figure 7 11 Cylinder mounting space required

In some cases, the customers would find that the available space for fitting the cylinder in the machine is less than the stroke length required for the working of the cylinder.

In this type of application requirements, it is common to look for a telescopic cylinder which can provide a longer stroke length

Telescopic cylinders are a special design of a hydraulic cylinder or pneumatic cylinder which provide an exceptionally long output travel from a very compact retracted length. Typically, the collapsed length of a telescopic cylinder is 20 to 40% of the fully extended length depending on the number of stages. Some pneumatic telescoping units are manufactured with retracted lengths of under 15% of overall extended unit length.

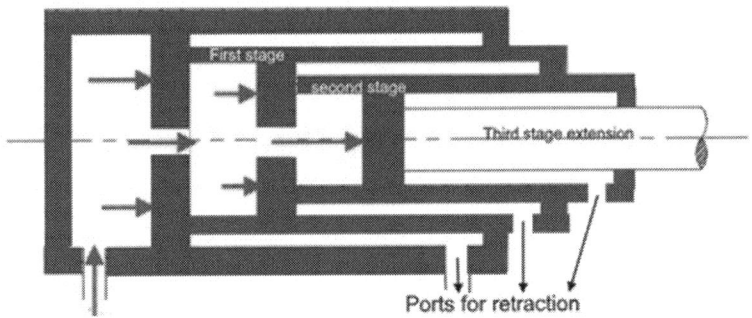

Figure 7 12 Telescopic cylinder

Please refer Figure 7 12.A telescopic cylinder with three stages is shown. The red arrows indicate the oil pressures on first stage, second stage and third stage of the cylinder. To retract the cylinder ports shown should be used.

7.3 POINTS TO RECALL

1. The most popular usage is for double acting hydraulic cylinders.
2. The student should understand the terminologies of full bore area(Piston area), annulus area, stroke length, cylinder length, pressure, and load.
3. Applications of single acting cylinders, and telescopic cylinders should be clear to the students.

8. HYDRAULUC ACTUATORS-2

8.1 LEARNING OUTCOMES

By the end of this chapter, the student will understand,

1. The types of Hydraulic motors
2. The specifications/ parameters of hydraulic motors.

8.2 HYDRAULIC MOTORS

A hydraulic motor is a mechanical actuator that converts hydraulic pressure and flow into torque and rotation. The hydraulic motor is the rotary counterpart of the hydraulic cylinder. We can use hydraulic motors for many applications, such as winches, crane drives, excavators, mixer and agitator drives, roll mills, etc.

The power produced by a hydraulic motor is determined by the flow and pressure drop of the motor. The displacement and pressure drop of the motor determines the torque it generates. The power output is thus directly proportional to the speed. The motors range from high speed motors of up to 10,000 rpm to low speed motors with a minimum of 0.5 rpm. Note that low speed hydraulic motors are designed in such a way that large torques are generated at low speeds. High speed motors have better operational characteristics at speeds that are at least higher than 500 rpm. Hydraulic motor comparison

8.3 HYDRAULIC MOTOR COMPARISON WITH ELECTRIC AND PNEUMATIC MOTORS

S.No	Electric motors	Hydraulic motor	Pneumatic motor
1	Least expensive	Expensive	Moderate investment
2	Preferred in high revolutions per minute- 100 RPM and above	Preferred in low speeds – even less than 100 RPM applications	Better suited for less than 1000 RPM applications
3	Low to medium Torque	High Torque	Low torque
4	Over loading causes excessive damage	Overloading just stalls the motor	Overloading only stalls the motor.
5	Not preferable for underwater /dust/dirty applications as insulation/design issues	Properly designed hydraulic motor highly suitable for underwater applications or applications involving dirt/dust etc.,	More suitable than electric motor.
6	Speed control requires Additional circuits	Speed control is simple with flow control valves	Flow control is simple with Flow control valves

Figure 8 1 Hydraulic motor comparison

8.4 TYPES OF HYDRAULIC MOTORS AND SELECTION OF MOTORS

The different types of hydraulic motors are given as below.

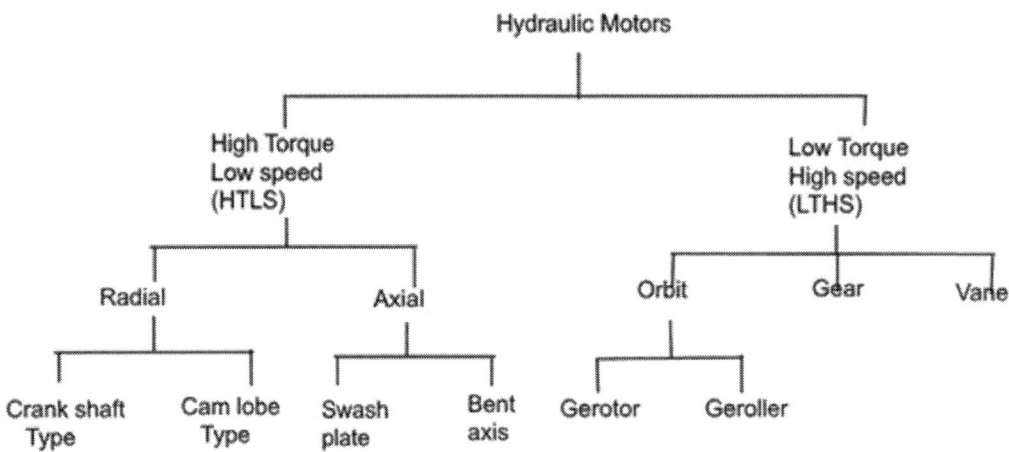

Figure 8 2 Types of hydraulic motors

With the above classification in mind, we first has to decide if the motor is for High torque low speed application or foe low torque high speed application.

LTHS(Low Torque High Speed) Opted in case of Moving heavy loads, like cranes winches etc.,

HSLT(High Speed Low Torque) Opted in case of steady loads moving at high speeds like fans, special purpose machine drives etc.,

High speeds are generally above 1000 RPM. However, it is also the load that decides the choice of the type of motors.

In addition to the torque/speed requirements following parameters are also to be considered:

1. Work Cycle – How long (i.e) at what speed and torque range the motor is to operate.
2. Working Atmosphere-Heat, dust level, vibration etc.,
3. Working Pressure
4. Displacement (Cubic cm/Revolution)

Manufacturer' catalogue would normally cover details of 3 and 4 for each type of motor.

The displacement of hydraulic motor is a main criterion for the selection of motors. It is the amount of oil required to turn the motor output shaft by one revolution. The displacement is expressed in terms of cm3/ revolution.

Torque is the force component of motor output. It is defined as the turning or twisting effort. The torque at the motor shaft is load multiplied by shaft radius of the motor.

Torque (Nt-m) = [Pressure(Bar) X Displacement(cm3/rev)]/62.83

Torque(Nt-m) = [Flow(l/min) X Pressure(Bar)]X 15.91/RPM

Torque(Nt-m) = [Power(KW) X9543]/RPM

Flow(l/min) = RPM X Displacement(cm3/rev)]/1000

From the above formulae, we can obtain following:

D = DISPLACEMENT (cm³/rev)   FLOW = Q (Litres/minute)

TORQUE = T (Nt-m)   PRESSURE = P (bar)

RPM = (revolutions/minute)

$$P\,(bar) = \frac{T\,(Nt\text{-}m) \times 62.83}{D\,(cm^3/rev)} \qquad Q\,(Litres/minute) = \frac{RPM \times D\,(cm^3/rev)}{1000}$$

$$T\,(Nt\text{-}m) = \frac{P\,(bar) \times D\,(cm^3/rev)}{62.83} \qquad RPM = \frac{1000 \times Q\,(Litres/minute)}{D\,(cm^3/rev)}$$

$$D\,(cm^3/rev) = \frac{T\,(Nt\text{-}m) \times 62.83}{P\,(bar)} \qquad D\,(cm^3/rev) = \frac{Q\,(Litres/minute) \times 1000}{RPM}$$

Figure 8 3 Hydraulic Motor Formulae

[Please refer the appendix 2 for FPS units]It is emphasized here that the users check the manufacturer's catalogue for torque characteristics and identify the higher torques needed in one particular cycle and accordingly choose the motor.

In addition, following points also to be considered while selecting the motor.

1. If the hydraulic motor is on Continuous working or intermittent or occasional duty.

2. The sequence of pressure/speeds at which the motor is expected to operate

3. The length of time of each phase of the cycle

4. the frequency of cycles during the work period.

5. working atmosphere.

8.5 POPULAR TYPES OF HYDRAULIC MOTORS

8.51 Gear Motor

A gear motor (external gear) consists of two gears, the driven gear (attached to the output shaft by way of a key, etc.) and the idler gear. High pressure oil is ported into one side of the gears, where it flows around the periphery of the gears, between the gear tips and the wall housings in which it resides, to the outlet port. The gears then mesh, not allowing the oil from the outlet side to flow back to the inlet side.

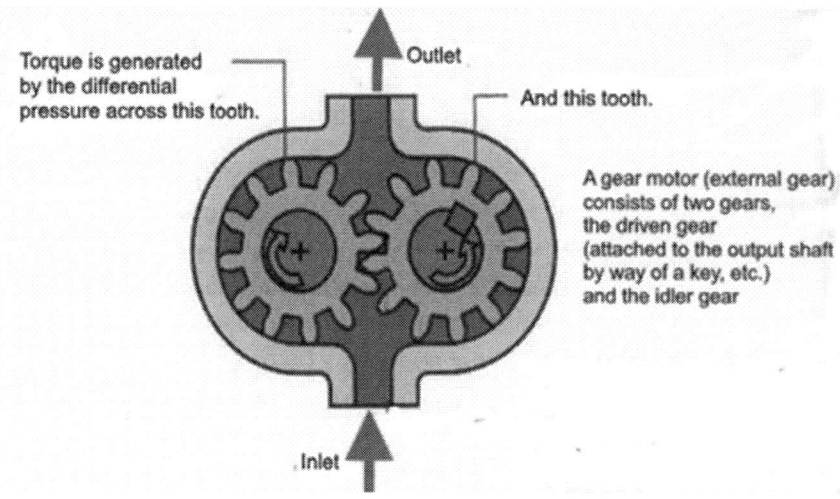

Figure 8 4 Gear Motor

An especially positive attribute of the gear motor is that sudden breakdown is less common than in most other types of hydraulic motors.

This is because the gears gradually wear down the housing and/or main bushings, reducing the volumetric efficiency of the motor gradually until it is all but useless. This often happens long before wear causes the unit to seize or break down.

8.5.2 Vane Motor

This particular motor provides rotation in only one direction. The rotating element is a slotted rotor which is mounted on a drive shaft. Each slot of the rotor is fitted with a freely sliding rectangular vane. The rotor and vanes are enclosed in the housing, the inner surface of which is offset from the drive shaft axis. When the rotor is in motion, the vanes tend to slide outward due to centrifugal force. The distance the vanes slide is limited by the shape of the housing.

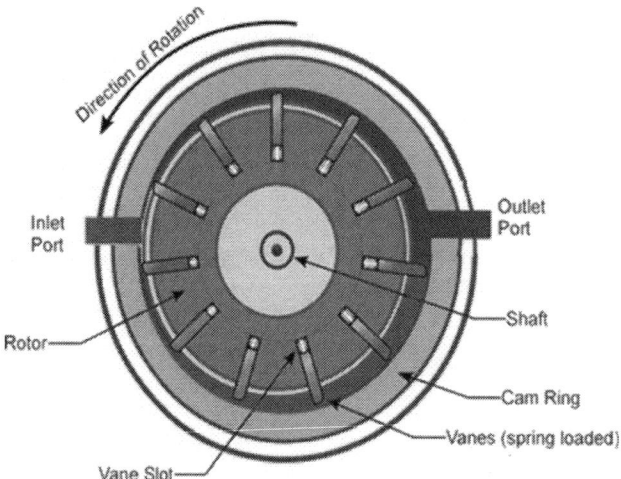

Figure 8 5 Vane Motor

In the above Motor, the rotor is eccentrically located inside the am ring space. As in vane pump application, even, in Hydraulic vane motors, we can have an elliptical cam ring construction to balance the forces.

The vane motor can have a construction as shown below. This is similar to Vane pump construction and the need for having an elliptical cam ring to balance the forces.

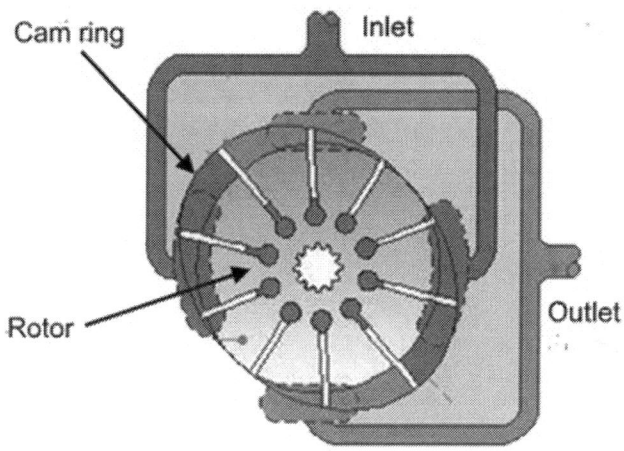

Figure 8 6 Balanced design vane pump

We have seen a few high speed low torque motors. Some examples of high torque low speed motors constructions/ principles are given below.

8.5.3 Piston motors

As in pumps, these motors are available in 3 types of constructions.

1. Axial or inline piston motors

2. Radial piston motors

3. Bent axis type motors. We have discussed in chapter 6, pumps of above configurations. Here in motors, the reverse of the process takes place. That is, the pressurized oil acts on the pistons to impart and translate the torque in to rotation.

Piston type motors find applications in mobile hydraulics like in winches and excavators or where ever heavy loads are to be handled at low RPMs.

8.6 DISADVANTAGES OF HYDRAULIC MOTORS

1. Hydraulic motors require a hydraulic power unit. This means we must have hydraulic pump driven by a prime mover so that the pumped oil can be used to drive the hydraulic motor. This kind of set up increases the cost.

2. Hydraulic motor speed sometimes need a reduction gear to further reduce the speed. This would also mean an increase in torque at the output shaft of the reduction gear. Design has to take care of this increase in torque while selecting the motor.

8.7 HYDRAULIC CIRCUIT SYMBOLS FOR HYDRAULIC MOTORS

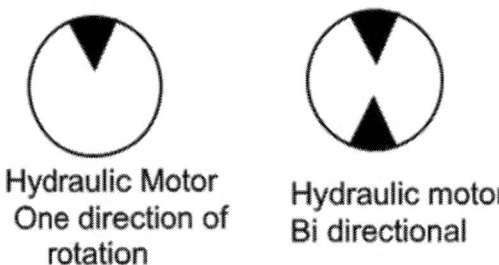

Figure 8 7 Hydraulic motor circuit symbols

8.8 POINTS TO RECALL

1. The students should be familiar with the different types of motors construction and their application.

2. It is important that the student remember the formulae associated with torque/ flow calculations for hydraulic motors.

3. Understand that the starting point for motor selection is the load/Torque characteristics, needed for the application.

9. HYDRAULIC ELEMENTS

9.1 LEARNING OUTCOMES

By the end of this chapter the student will understand

1. The different elements that are most frequently used in a hydraulic system(Hydraulic circuit) and their functions.

2. The fundamentals of building a hydraulic circuit.

9.2 HYDRAULIC ELEMENTS AND ACCESSORIES

The different components of a hydraulic system can be classified in to following groups.

9.2.1 Hydraulic Control elements

These are the valves that control the flow/ direction and pressure. The classification is as follows.

Control element	Function	Major types
Direction Control Valves	To change the direction of oil flow into the actuators (cylinders/ Motors) or into a branch of the circuit.	These are classified by the number of ways and number of positions of the valve.
Pressure Control Valve	To control the pressure of oil in the system.	1. Pressure relief valve 2. Pressure reducing valve 3. Pressure sequence valve
Flow Control Valve	To adjust the flow rate of oil after pumping the oil into the system. The speed of the actuator depends on the flow rate.	1. Flow control valve without reverse free flow 2. Flow control valve eith reverse free flow

Table 9 1 Types of control elements

9.2.2 Hydraulic accessories

These are the components that are used to store, filter the oil, measure pressure or indicate the oil level. Further, there is an accessory known as Accumulators which are used for storing high pressure oil.

Hydraulic accessories	Function	Types
Reservoir	To store oil- Dissipates heat of oil.	Mainly size dependent-The size is liters/ gallons based.
Breather filler	To fill the oil into the reservoir. Also, facilitates breathing- Oil is pumped out and the level in the reservoir goes down- when the oil returns after work the level goes up. The air is drawn into the tank when the oil level goes down and the air is sent out when oil comes back into the reservoir.	Available in sizes based on air flow rate.
Coupling	To connect the pump shaft with prime over shaft(Motor/ Engine).They transfer the turning moment generated by the motor to the pump.	There are different types of couplings based on its construction. Eg Rubber couplings, Gear couplings, Flange couplings etc.,
Oil level gauge / Indicator	Used for monitoring fluid levels and temperature in various types of reservoirs.	Based on size, Mounting and $^{\circ}$C or $^{\circ}$F indication
Strainers	This is used inside the reservoir on the suction pipe to ensure bolts/ nuts or parts do not get into suction side of the pump. They are intended to keep larger solid contaminants from entering the hydraulic system. A drawback is that they are quite inaccessible for service and cleaning. If they become restricted due to excess contamination, they can cause cavitation and damage to system pumps	Depends on the mesh size used for straining and nominal flow.
Return line Filters	Mainly used for filtering the oil- Controlling contamination of oil in the system as the oil can easily get contaminated during usage by metal chips/ particles/dirt/paint/sealing parts etc.,- It is the task of the filter to reduce this contamination to an acceptable level in order to protect the components from excessive wear.	Rated by pressure, flow rate, and grade of filtration levels. Filters are available normally above 5 microns.

Figure 9 1 Hydraulic accessories

Further features of the hydraulic elements and the hydraulic accessories will be brought out as we build hydraulic circuits step by step in the chapters that follow.

9.3 POINTS TO RECALL

1. The student should understand the basic difference between hydraulic elements and the accessories.

2. The hydraulic elements are the main valves used in the hydraulic system.

3. Acessories has the function of supporting/ monitoring the system.

10. UNDERSTANDING HYDRAULIC CIRCUITS

10.1 LEARNING OUTCOMES

By the end of this chapter the student will be able to understand

1. The first step in making the hydraulic circuit. That is having a reservoir with all its accessories and up to Pressure relief valve.

2. The working of Pressure relief valve, Pressure reducing valve and Pressure sequence valve.

10.2 FIRST STEP IN THE PROCESS OF BUILDING A HYDRAULIC CIRCUIT

1. The first step in a hydraulic circuit is to represent a reservoir that stores hydraulic oil for pumping out to the system elements and to the actuators.

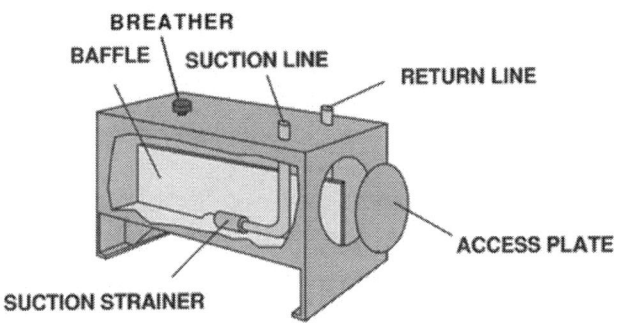

Figure 10 1 Hydraulic oil Reservoir

Please refer Figure 10 1. The traditional thumb rule is that the reservoir is sized 3 times the pump capacity. For instance if the pump is to deliver 30l/min(Fixed displacement pump), then the reservoir can be designed for a capacity of 100 liters of oil. If the system is with a variable-displacement pump mean flow rate can be considered for calculating the reservoir size.

The top surface can be used for mounting the pump coupled with a motor. It is possible to have a pump inside the tank immersed in oil and the motor is mounted vertically on the surface of the tank and coupled with the submerged pump inside.

The reservoir should contain additional space equal to at least 10% of its fluid capacity. This allows for thermal expansion of the fluid and gravity drain-back during shutdown, yet still provides a free fluid surface for deaeration.

Systems exposed to high ambient temperatures require a larger reservoir unless they incorporate a heat exchanger. This heat is generated when the hydraulic system produces more power than is consumed by the load. A system operating for significant periods with pressurized fluid passing over a relief valve is a common example.

A baffle plate is provided normally to separate the suction side of the pump from the return line side of the system. This helps in reducing the turbulence inside the tank. The student can refer to Figure 5 1 (Though Baffle plate is not clearly seen in this figure).

The first step in building the hydraulic circuit is to be clear about the reservoir and the accessories that go with it as shown

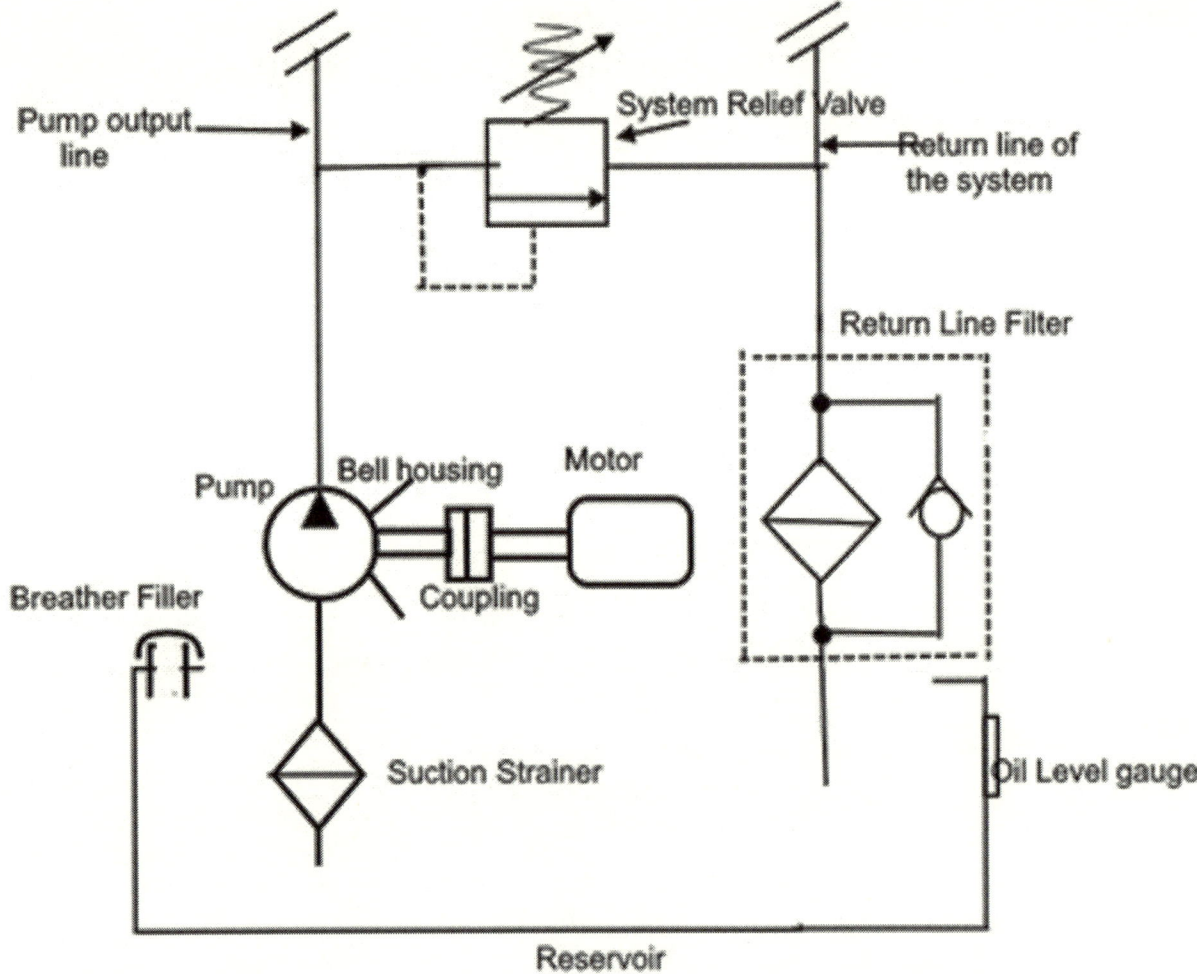

Figure 10 2 Reservoir and the accessories(Reservoir circuit)

In the Figure 10 2, the reservoir, its accessories, pump, Motor and the system pressure relief valve is shown. It is important that the student understands that the system relief valve is always connected between the pump output line(pressure line) and the return line of the system.

10.2.1 System pressure relief Valve

The Pressure relief valve in a hydraulic system is similar in function to that of a fuse or circuit breaker in an electric circuit. An electric circuit never blows a fuse unless it is overloaded.

In the same manner, system relief valve protects the hydraulic system. It is set at a pressure higher than the system operating pressure. Therefore, when the system operating pressure is exceeded the relief valve opens and allows the pressure line to be connected directly to the tank line. When this excess pressurized pump flow goes to the tank, it generates heat. It is common to say that the system pressure blows over the relief valve. The working principle of system relief valve is given below

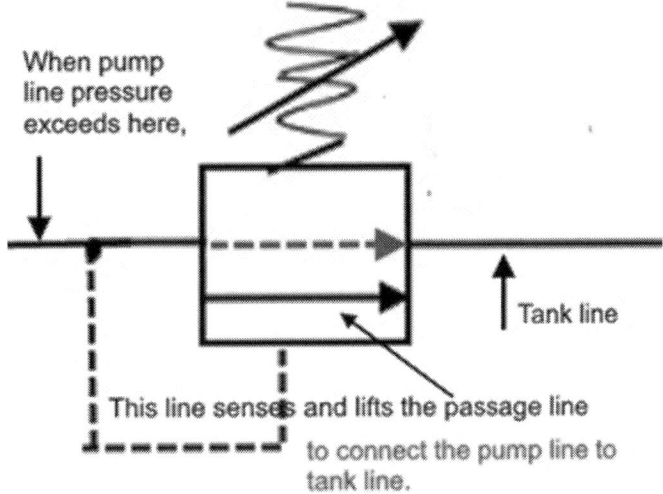

Figure 10 3 System relief valve working principle

The one shown in Figure 10 3 is referred as direct acting relief valve.

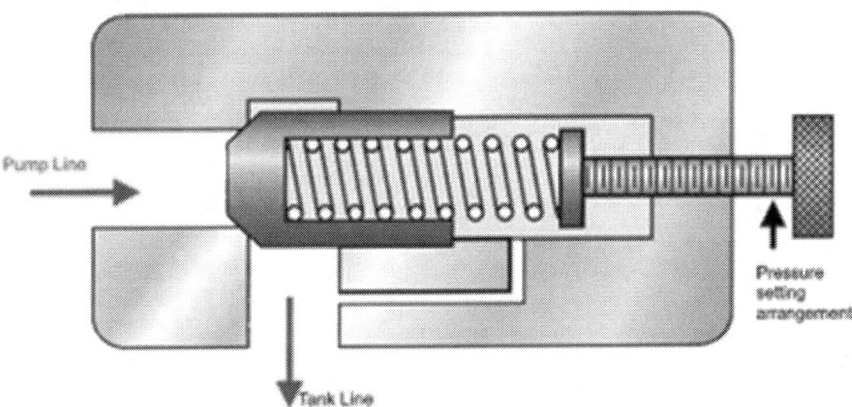

Figure 10 4 Constructional features of direct operating relief valve

Generally, the direct acting relief valve is used up to 50l/min flow or less. For flows upto 400 l/min , the normal practice is to go in for pilot operated relief valve(Also referred as compound relief. This valve has two stages as shown in Figure 10 5

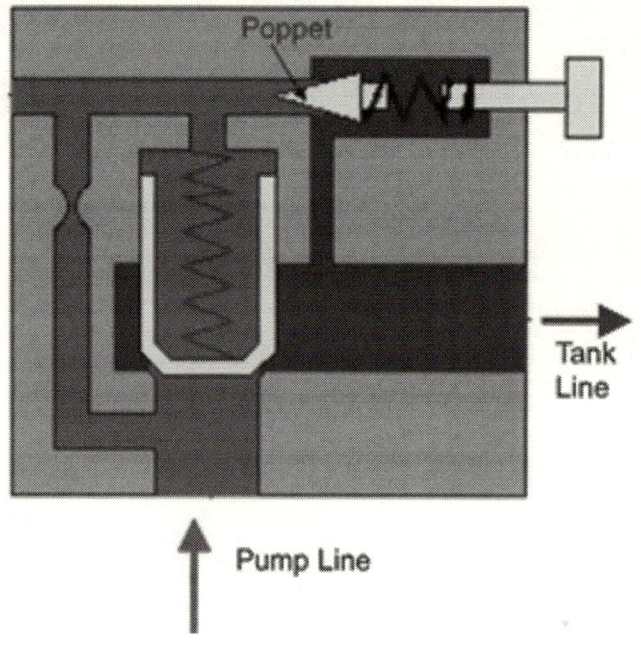

Figure 10 5 Pilot operated Pressure relief valve constructional features

In the above figure the pressure is sensed by the relief valve (Horizontal kept above the main vertical relief valve) and when the set pressure is exceeded it allows a partial opening of the poppet and the pressurized oil flows to the tank. This flow changes the pressure differential across the main(vertical) relief valve and this makes the valve lift its seat to pass larger flow of pressurized oil to the tank line.

Cracking pressure and pressure override —The pressure at which a relief valve first opens to allow fluid to flow through is known as cracking pressure. When the valve is bypassing its full rated flow, it is in a state of full-flow pressure. The difference between full-flow and cracking pressure is sometimes known as pressure differential, also known as pressure override.

10.2.2 Pressure Reducing valve

While on the subject of pressure relief valve, let us also understand other family of pressure control valves. Pressure reducing valve is used when only one branch circuit requires lesser pressure compared to the operating pressure of the rest of the system. Pressure reducing valves control their outlet or downstream pressure only and not the system pressure. This is valve is required in multiple actuator circuits, where it is impossible to size all actuators to operate at maximum system pressure. It should also be noted that in comparison with Pressure relief valve(which is normally closed), this valve is normally open.

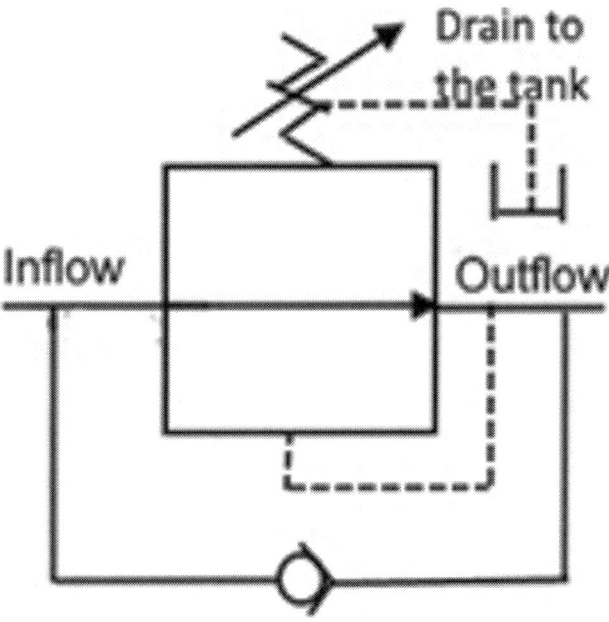

Figure 10 6 Hydraulic symbol of Pressure Reducing valve

As mentioned, a pressure-reducing valve does not allow pressure downstream of the valve to exceed the set point. Suppose a workpiece must be clamped and too much clamping force will damage the workpiece. In such an application, a pressure-reducing valve is used to limit clamping pressure. This valve is externally drained to the tank as shown in Figure 10 6

10.2.3 Pressure sequence valve

Pressure sequence valve is very much similar to pressure reducing valve and hence it is appropriate to understand the pressure sequence valve at this stage. Both this valve and the pressure reducing valve are externally drained to the tank as shown in the symbols.

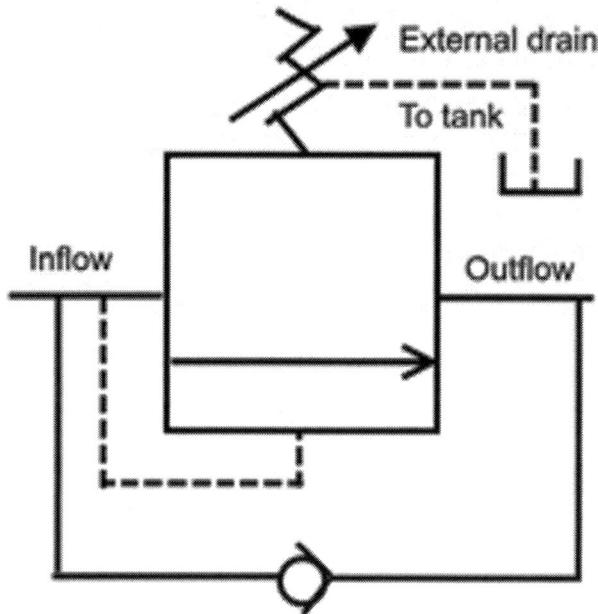

Figure 10 7 Hydraulic symbol of Pressure sequence valve

.The sequence valve is used to ensure that a certain pressure level is achieved in one branch of the circuit before a second step or a sequence is actuated. Example is holding (clamping a work piece) for machining. First we have to ensure that pressure is applied so that the work piece is held properly. Then, the next sequence of action is the work piece is extended to make contact with the cutting tool

If the first action is not done, and the second sequential action is performed, the work piece might fall off – which is not desirable.

Pressure sequence valve, in this example would ensure a set pressure for clamping is applied and then only it opens for the next action.

10.3 POINTS TO RECALL

1. The students should remember that the system relief valve is always connected between the pump line and the tank line.

2. If the oil blows over relief valve to the tank, due to safety considerations, then the oil pressure generates heat.

3. Pressure reducing valve is used when lesser pressure (than the operating pressure) is required to be maintained in a branch of the circuit.

4. Pressure sequence valve is used when a particular action involving maintenance of pressure is followed by the next sequential actuation.

We shall be looking at specific hydraulic circuits for even better understanding of these valves.

We shall not be drawing the reservoir circuit every time. It would remain as standard and the circuit above the reservoir circuit will only be shown. The student should be able to visualize the reservoir circuit with the circuit drawn in the chapters that follow.

11. UNDERSTANDING HYDRAULIC CIRCUITS-2

11.1 LEARNING OUTCOMES

By the end of this chapter the student will understand

1. Connection of hydraulic circuit beyond the pressure relief valve.(Beyond the reservoir circuit)

2. The types and function of Direction control valves.

11.2 SIZING OF HYDRAULIC CYLINDER, PUMP AND THE MOTOR FOR HYDRAULIC CIRCUIT.

In the last chapter, we have developed a circuit up to and including pressure relief valve. Let us consider a simple circuit having only one hydraulic cylinder in the system.

Let us say that the cylinder has to lift a load of 1 ton (1000 Kg) as shown below.

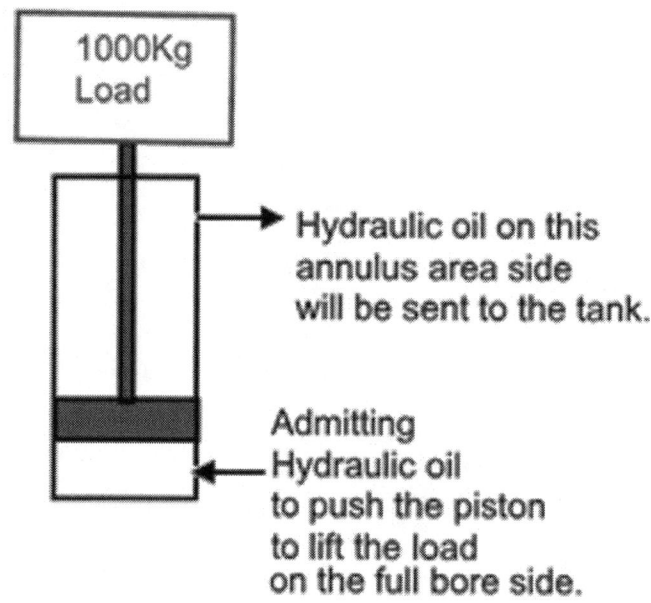

Figure 11 1 Hydraulic cylinder to lift a load of 1000 Kg(1T)

We are aware of the formula P(Kg/Sq.cm) = Force(Kg) / Area(cm2)

By looking at a hydraulic cylinder manufacturers guide, we see that the manufacturers have standard bore(D) and rod diameters(d) for the cylinders. Let us for select a cylinder bore diameter of 40 mm.

The full-bore area of the cylinder is $(\pi D^2)/4 = [\pi * (4^2)]/4 = 12.56 cm^2$

Pressure Kg/cm2 = 1000 kg/12.56cm2 = 79.62 Kg/cm2

We would need a pump that has the capacity to develop a pressure of 80 Kg/cm2

In case we choose a pump that can develop a maximum pressure (say) 120 Kg/cm2 and delivers (say) 10 l/min.

The flow rate 10l/min = Q= A(area)* V(velocity)

10 l/min = 12.56 cm2 * V

(1 liter = 1000cc)

(10000 cc /m)/ 12.56 cm2

796 cm/min = V

(796 cm)/ 60 = v = 13.2 cm/second

It is possible to increase the speed by choosing a higher flow rate of the pump.

To recapitulate,

1. The hydraulic designer should first know the load to be tackled by the hydraulic system.

2. A hydraulic cylinder(Let us consider a double acting cylinder) of suitable bore diameter should be chosen from any suitable manufacturer's catalogue.

3. The pressure required to do the job can be calculated using the standard formula of P = F/A.

4. For a simple lifting job as above a fixed displacement vane type pump can be chosen considering the cost aspect.

5. The velocity of lifting the load(speed of movement of the cylinder) can be calculated using the formula Q = A*V

7. For working out the stroke length, we have to take into account the height to which the load is to be lifted the lifting height is 1 meter, then the stroke length can be 1 meter.

8. we have to then calculate the motor KW required to drive the pump. The best way to assess the motor KW requirement is from the performance curves of any reputed manufacturer.

The performance curves of PV2R1-8 vane pumps manufactured by Yuken India Ltd.,is given in the Vane pump performance curves (The vane pump PV2R1-8 performance curve is given below.

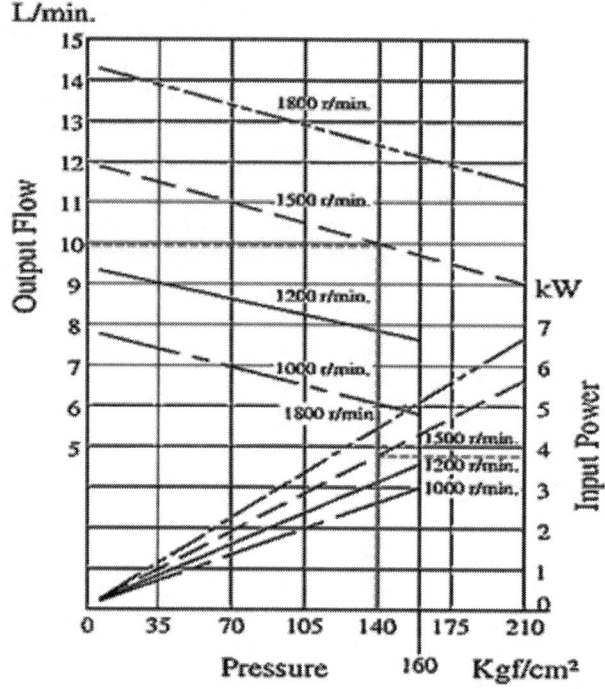

Figure 11 2 Vane pump performance curves(Yuken India Ltd.,)

The red dotted lines show the selected pump – capable of delivering 10l/min at 140 Kg/cm2 and the input power required is about 3.9 KW. From the motor manufacturers range of motors, perhaps a motor delivering 5KW at 1500 rpm can be chosen.

With the pump and the motor chosen, we can now go ahead to build the circuit.

We have already built up the circuit up to the reservoir including the pressure relief valve. We just have to include in the circuit the hydraulic cylinder and a suitable direction control valve.

11.3 FUNCTION OF DIRECTION CONTROL VALVE

The function of direction control valve is to admit hydraulic oil in to the ports of the double acting cylinder. It should admit oil through one port (Port 1) and the cylinder piston rod gets pushed by the pressurized oil. At the same time oil on the other side of the piston (rod side- annulus area side) should be drained to the tank, through Port 2, so that the piston can move encountering only the load pressure.

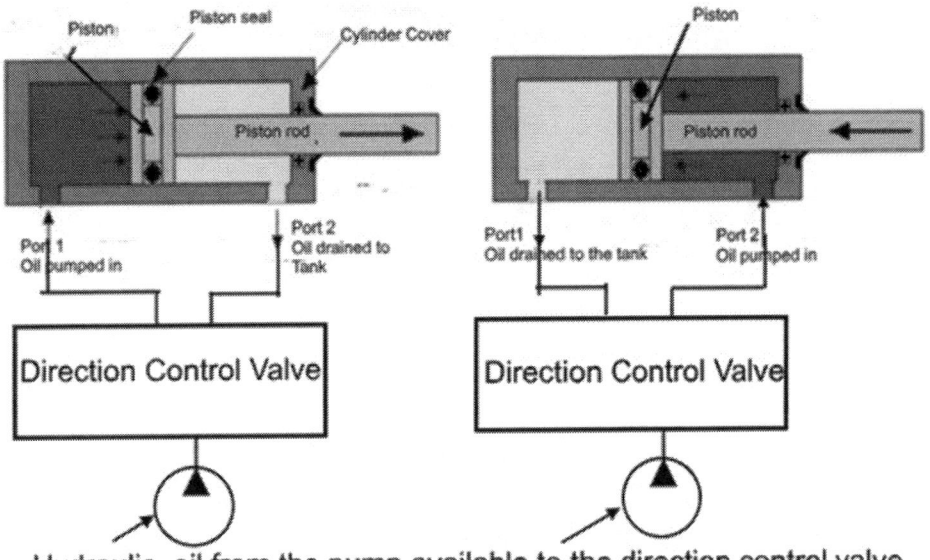

Figure 11 3 Hydraulic cylinder with Direction control valve

Then, once the piston reaches the other extreme end, the oil direction in to the cylinder should change. Oil should now enter port 2 (Rod side) and the piston should retract. The oil that got filled up on the full-bore side should be drained to the tank.

In the above Figure 11 3, Oil under pressure is shown in red. Oil to be drained is shown in yellow. The direction control valve is shown as a rectangle with oil from the pump getting in to the DC valve (Direction Control herein after referred as DC).

From the above Figure 11 3, it should be clear that at one time the direction of oil inflow should be towards port1 and the out flow of oil should be from port 2. The same DC valve then should admit oil from port 2 and facilitate draining from port 1.

This change of direction of flow of oil is accomplished by the direction control valve.

For the sake of building the circuit we are going to consider a direction control valve as shown.

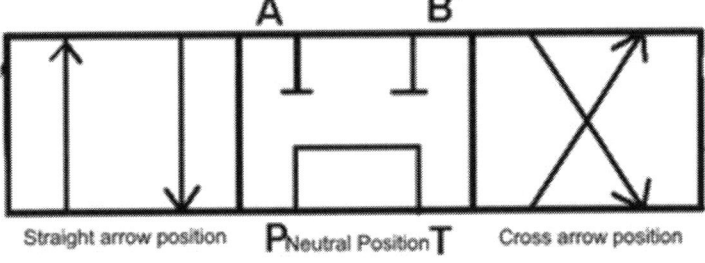

Figure 11 4 Spool of direction control valve

The Figure 11 4 shown is referred as the spool of the direction control valve. It has three positions. The first position is straight arrow (straight passages) The middle one is the neutral position of the valve and the third position is the cross arrow(cross passages) position. While connecting the pump line it has to be with passage marked P. Here in this spool configuration P and T are connected.

11.4 A SIMPLE HYDRAULIC CIRCUIT -CONNECTING DIRECTION CONTROL VALVE TO THE CYLINDER

The two ports of the cylinder should be connected to the ports. Port 1 to A and Port 2 to B. The cylinder connections with the DC valve is shown below.

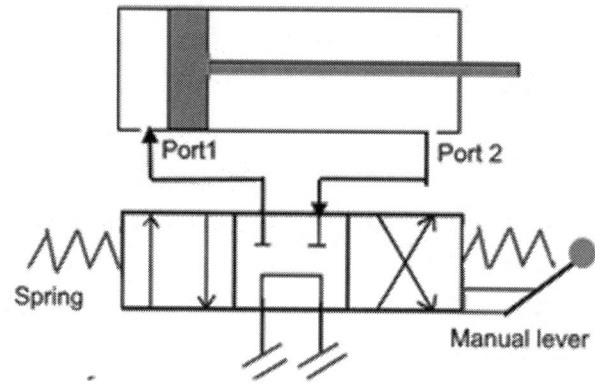

Figure 11 5 Cylinder with DC valve - Manual

The DC valve shows a spring on the left side and a spring with manual lever on the other side in Figure 11 5.

Now, we can connect the above with the reservoir circuit made by us earlier.

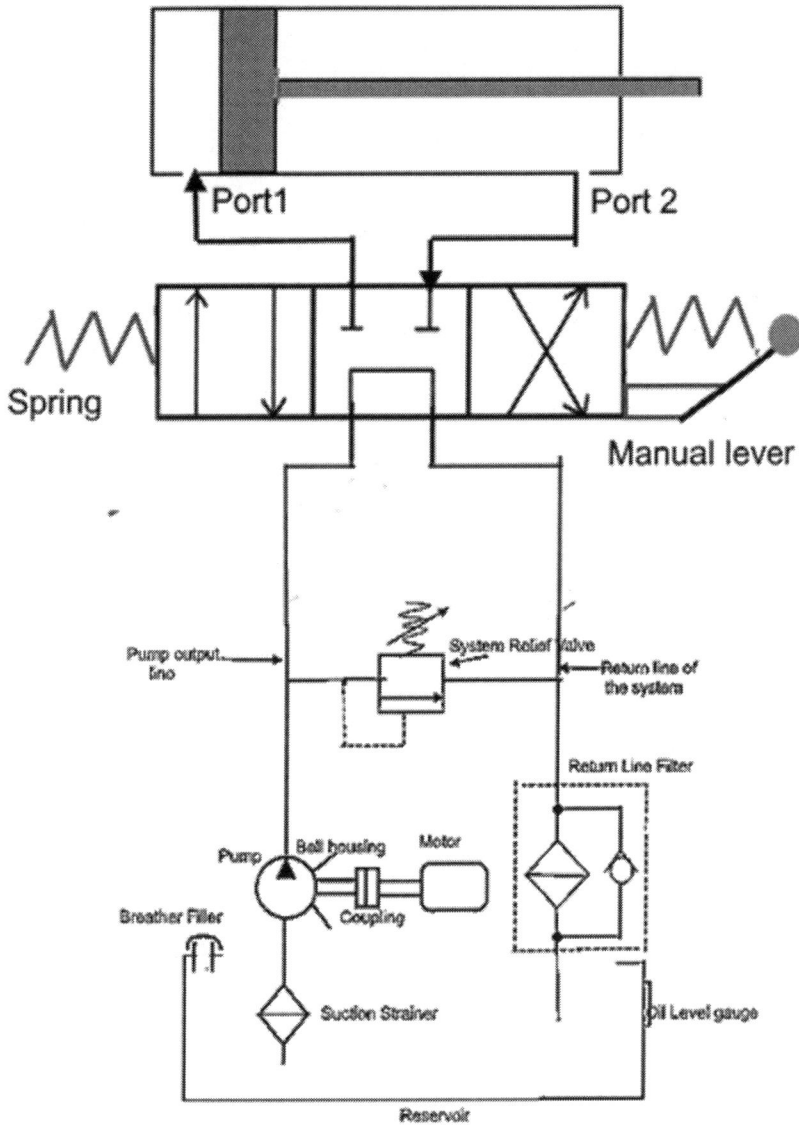

Figure 11 6 Simple hydraulic circuit

The DC valve shown in the Figure 11 6, shows a spring on the left side and a spring and a manual lever on the right side. In the initial stage, the valve position is kept neutral because of the springs on either side. In this position the oil from the pump is connected to the tank line through the DC valve and the oil gets drained. When the lever is moved to the left, the first position is activated and the pump line gets connected to the first passage. (marked with upward arrow) and the oil gets in to port1 and pushes the piston forward.

The port2 side gets connected to the down ward straight arrow position and then drained to the tank through the DC valve.

In the example, we are considering the load gets lifted by the stroke length. If now the lever is moved to the neutral position, both the cylinder ports are blocked in the DC Valve and the load is locked in lifted position.

However, if the DC valve lever is moved to the extreme right position then the pump line gets connected to the cross-port passage and the oil flow is now to port 2 of the cylinder. The piston moves retracting and the other cross port (downward arrow) gets connected to port 1 and the oil gets drained from this side to the tank through the DC valve.

A very simple arrangement of hydraulic circuit is covered in this chapter. We shall further hydraulic circuits after understanding the different types of direction and flow control valves.

11.5 POINTS TO RECALL

In this chapter the student the student should have understood.

    1. The connections between the double acting cylinder and the direction control valve.

    2. Connecting the DC valve and the cylinder set up to the reservoir with relief valve set up.

12. TYPES OF DIRECTION CONTROL VALVES AND THEIR ACTUATION

12.1 LEARNING OUT COMES

By the end of this chapter the student will understand the different types of direction control valves and their symbols.

12.2 DIRECTION CONTROL VALVE CLASSIFICATIONS

The DC valves can be classified according to following.

1. Number of positions – Two position and three position valves.
2. Spool configurations – The spool is made different types up of passages that are drilled in the valve body.
3. Actuation mechanism- These spools mentioned above have to be moved to connect with the pump and tank drain lines.

In the simple circuit in Figure 11 6, we have used a 3 position DC valve with manual lever for actuation. In three position DC valves, we always have a neutral position at the middle.

The positions on the left and right side of the DC valve generally remain same in all three-position type of valves. Only the configuration of the neutral positions change.

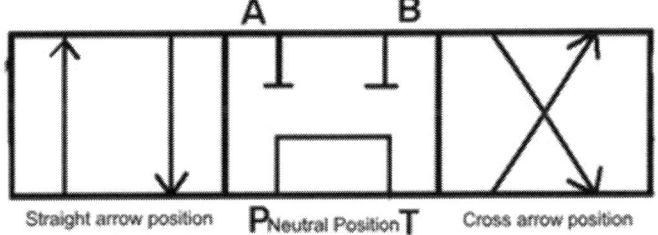

Figure 11 4 is reproduced above to further explain this point. The notations A and B are Port 1 and 2 respectively. In the next page, we are comprehensively covering images of all types of DC valves for better understanding and comparison.

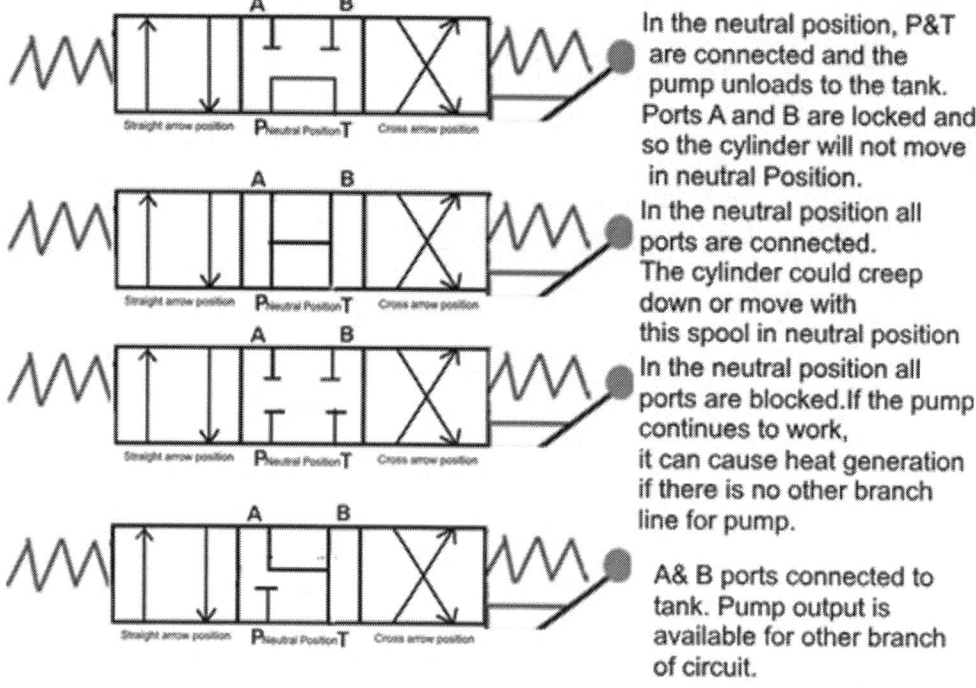

Figure 12 1 manual lever operated DC valve with different spools.

The figure above shows the spool configurations of four most popular spool types normally used in the industry in 3 position 4 way hydraulic valves.

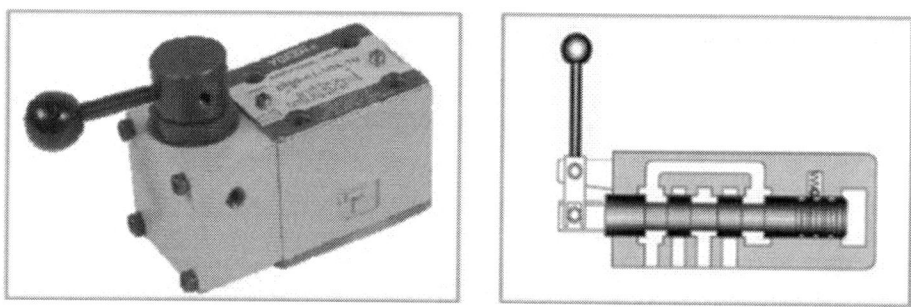

Figure 12 2 Manual lever operated DC valve Yuken India Ltd

Please refer Figure 12 3 ,In respect of number of positions, the student must by this time be clear. In terms 4 ways-To find out how many ways are available in a valve(In the case of 3 position valve)- consider the neutral position.

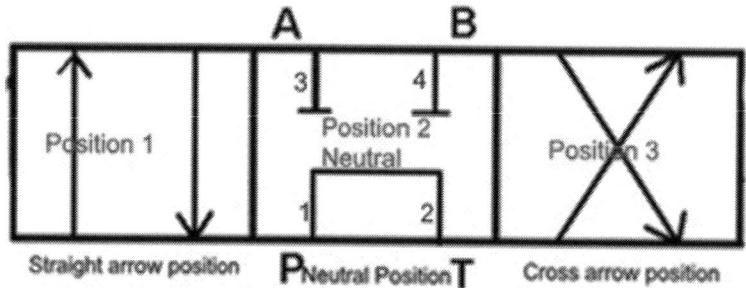

Figure 12 3 -3 position 4 way valve

In the neutral position, we have marked the number of ways as 1,2,3 and 4. In this case, ways 1 and two are connected and 3 and 4 are blocked. At times this valve is referred as 4/3 valve – or 4 ways 3 position, Tandem (P and T connected) DC valve. All the valves in Figure 11 7 are 4/3 valves.

These 4/3 valves also can be actuated by Solenoids. The Solenoids, which are electrical accessories, move(push) the position to connect with the pump and tank line.

The Solenoid ratings are available in 12V, 24V, 48V, 110 V and 220 Voltages. The 12V, 24V and 48 V are also available for direct current and 220 V for AC applications.

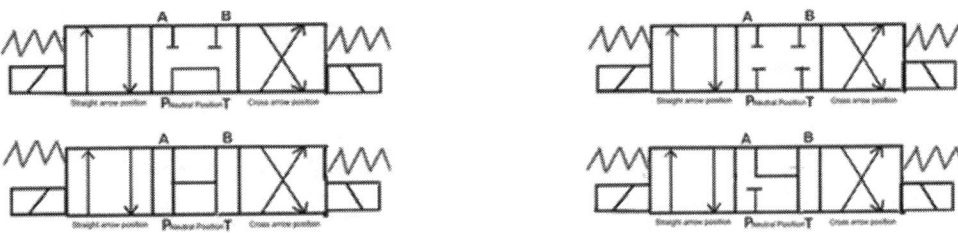

Figure 12 4 4way 3 position Solenoid operated valves

These valves also have springs on either side. This means that the pump line will be connected to the neutral position, if the Solenoids are not actuated. If the left side Solenoid is energized, then, the straight arrow position will get connected to the pump and the tank. Then, the connection to left Solenoid is cut off and the right Solenoid is connected. Then, the pump and he tank line of the cross-port positions get connected to pump and tank lines.

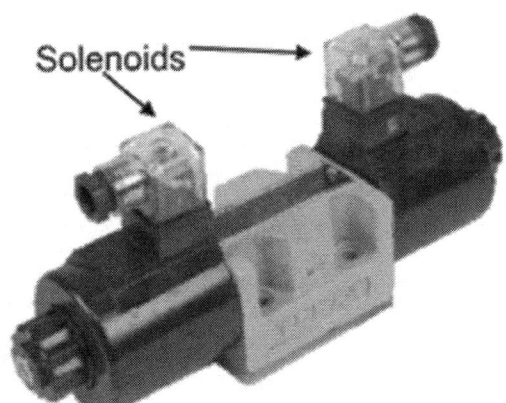

Figure 12 5 Double Solenoid DC valve manufactured by Yuken India Ltd.,

The same kind of logic and rationality is also applicable for two position DC valves. Generally, the most often used 2 position Dc valve spool valve is shown below.

Figure 12 6 4 way 2 position DC valve spool

The 4 ways mean the four entry/exit holes (ways) for the oil. Each arrow has one entry and one exit hole. Hence this valve is referred as 4 way 2 position valve or simply as 4/2 valve.

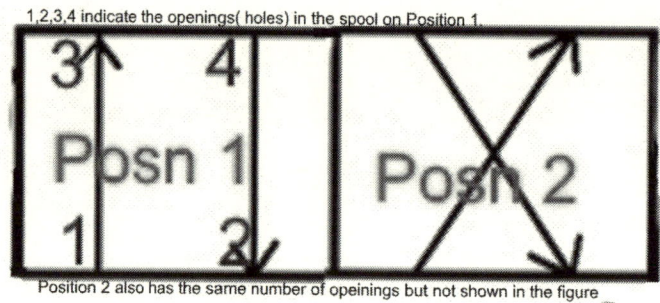

Figure 12 7 Single Solenoid 4 way 2 position Dc Valve

In, Figure 12 7 only Solenoid operated two position valves with different spool configurations are shown. Please note that manual operated two position valves do not find many applications.
In the two position valves shown, please note that the closest position to the spring is the normal resting position of the valve. In the first left extreme figure, it is the cross port position that gets connected when Solenoid is not energized. As and when the Solenoid is energized, the straight arrow position gets connected. If the connection to the Solenoid is off, then again cross port position is connected to pump and tank passages.

12.3 PILOT OPERATED DC VALVE

DC Solenoid operated is widely used for different applications.
The direction control valve is designed to handle particular rate of flow of oil passing through it. Normally, these direction control valves come in different sizes to handle 50l/min,100l/min,200 l/min 400 l/min etc.,In case of Solenoid operated DC valves, the Solenoid size (to move the position of the DC valve) is cost effective up to 100 l/min valve capacity. Beyond 100 l/min,the size of the Solenoid and the current required to move the spool become uneconomical and we have to think of a two stage valve.The main stage direction control valve is designed for large flows and over this main stage valve a Solenoid operated DC valve will be mounted.

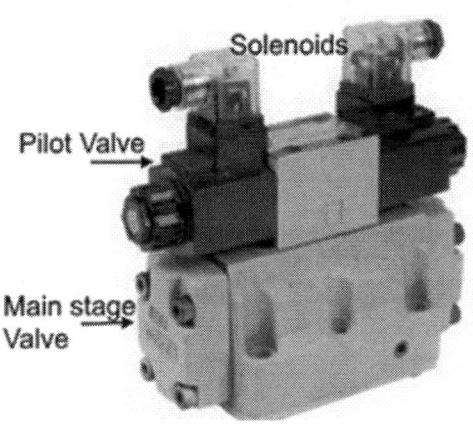

Figure 12 8 pilot operated DC valve of Yuken India Ltd.,

Please refer Figure 12 8. The principle here is that the pilot valve is connected to the pump flow when left side Solenoid is actuated. This oil flow is enough to push the spool of the main stage valve. Once the main stage spool is pushed, it makes way for the oil to flow to the actuator.

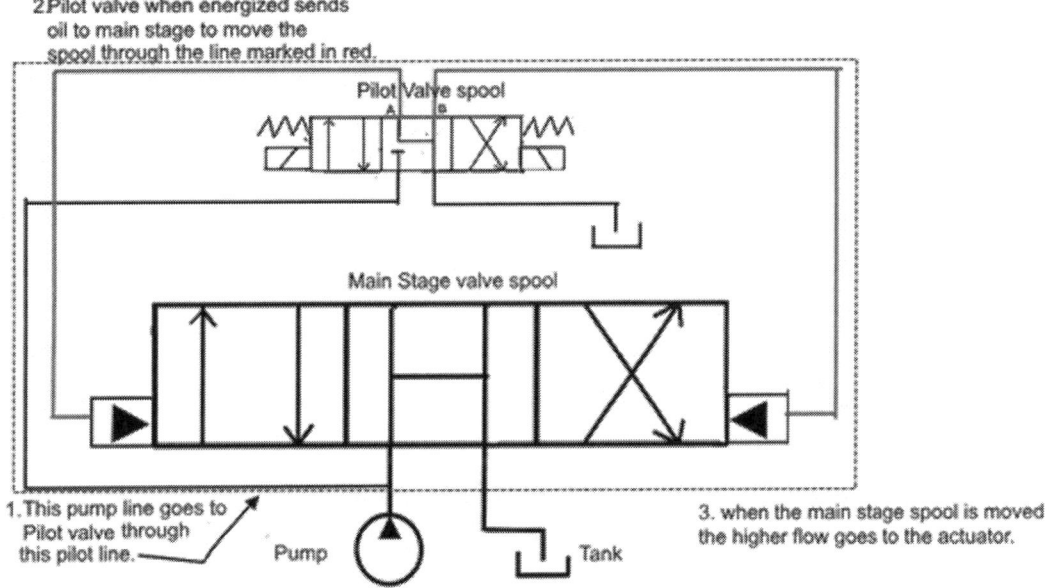

Figure 12 9 working principle of pilot operated DC valve.

The above Figure 12 9 is self-explanatory.

12.4 CHECK VALVE

Check valve is the simplest form of DC valve. These check valves allow oil to flow only in one direction.

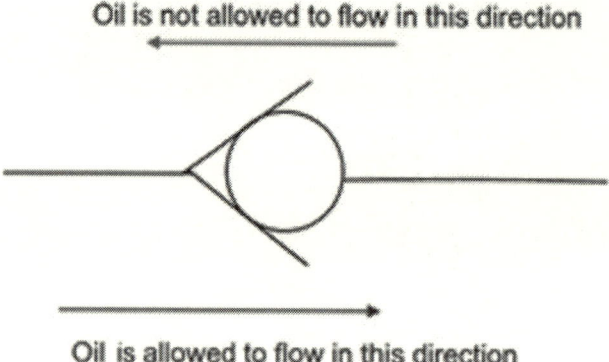

Figure 12 10 Symbol of check valve

Application wise, check valve is used, for example, on a pump delivery line so that oil from the pump goes to other hydraulic valves, but the oil is not allowed to flow back towards the pump.

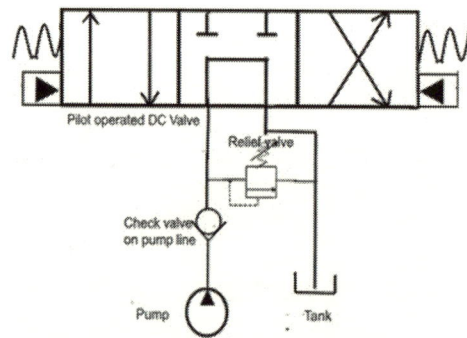

Figure 12 11 Check valve on pump line

It is also used in the tank line, when it is necessary to see that the oil return line is not completely empty by totally draining off to the tank. The only minor modification required for effective functioning of this type of check valve is that it has a light spring (of 0.5Bar) integrated along with it as shown in Figure 12 12

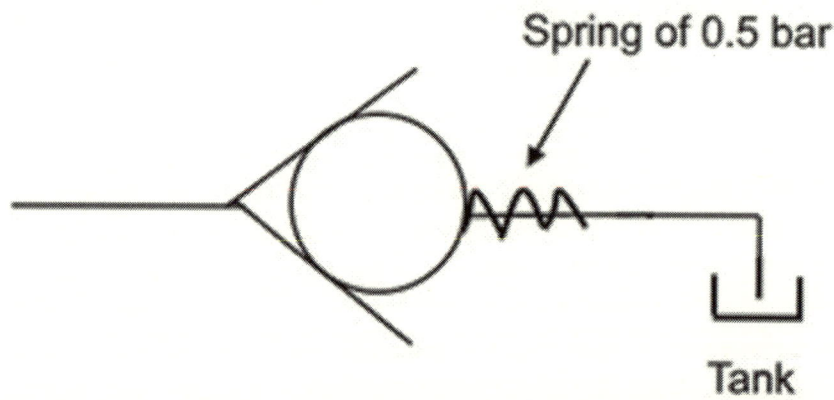

Figure 12 12 Check valve with spring

Following specifications relate to check valves:

1. The sizes of check valves -3/8", 1/2", 3/4", 1" and 1/4" BSP. These sizes can be directly threaded on to the pipe line.

2. Flow handling capacity of the check valve.

3. Flow handling capacity. For flow handling capacity of 100l/min, we can opt for right angle check valves which are designed for higher flow ratings. In this case the input and output ports are at right angles to each other.

4. Spring rating: This spring rating is applicable in the free flow direction. If a rating of 0.5 Bar is the spring rating, then the pressure in the direction of flow has to be more than 0.5 Bar for the oil to flow across the check valve.

5. The check valve comes either as pipe mounted or as stackable valve.

12.5 PILOT OPERATED CHECK VALVE

A pilot operated check valve is similar to a check valve with rated spring explained in the previous section. The application is explained in the Figure 12 13

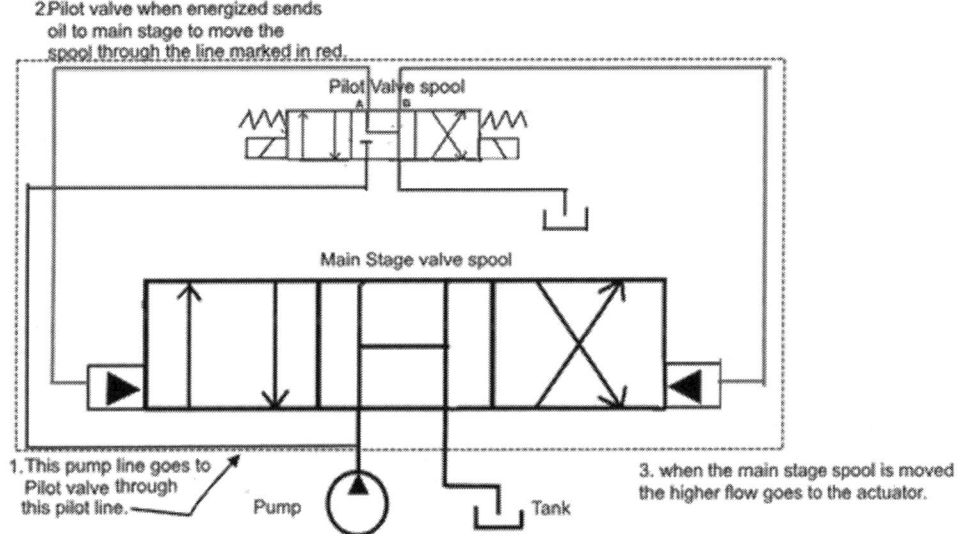

Figure 12 13 Application of pilot operated check valve

12.6 POINTS TO RECALL

In this chapter, the student should have understood

1. The types of DC valves based on positions, ways and spool configurations.

2. Application of pilot operated DC valves when the flow rate is more than 100 l/min.

3. Function and applications of check valves and pilot operated check valves.

13. UNDERSTANDING HYDRAULIC CIRCUITS-3

13.1 LEARNING OUTCOMES

By the end of the chapter, the student will understand,

1. Hydraulic circuits involving pressure reducing valve and pressure sequence valves and pilot operated check valves
2. Applications of flow control valves in the hydraulic circuits.

13.2 PRESSURE REDUCING VALVE & PRESSURE REDUCING VALVES IN A HYDRAULIC CIRCUIT

1. Let us first recall the function of a pressure reducing valve. This valve is used in a branch of a circuit where one has to maintain a pressure lower than the operating pressure of a system. This operating pressure is less than the system relief valve pressure.

2. The symbols of pressure relief valve and the pressure reducing valves are given below to refresh the memory.

Presssure Relief valve

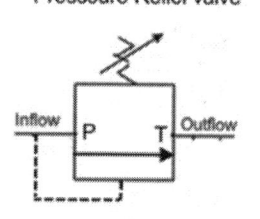

1. Sensing(dotted line) from pump line
2. The line is normally closed. Meaning P and T not connected. The bridge is below P&T passage line.
3. When the pressure goes beyond set pressure, the sensing line lifts the arrow to connect P and T and the pressured liquid flows to the tank.

Pressure Reducing Valve

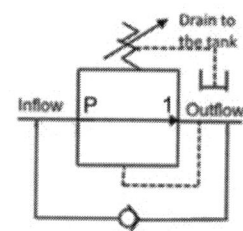

1. Sensing line is from outflow line unlike pressure relief valve.
2. The connecting line is normally open. It connects the pressure (pump line) to the port 1 line of cylinder.
3. The drain line is external so that outflow line is not disturbed.

Figure 13 1 Pressure relief valve Vs Pressure reducing valve

In addition to the points mentioned in Figure 13 1, please note that Pressure rlief valve is always connected between Pump(P) and Tank (T) lines.

The pressure reducing valve is connected close to the actuator which has to operate at a lesser pressure.

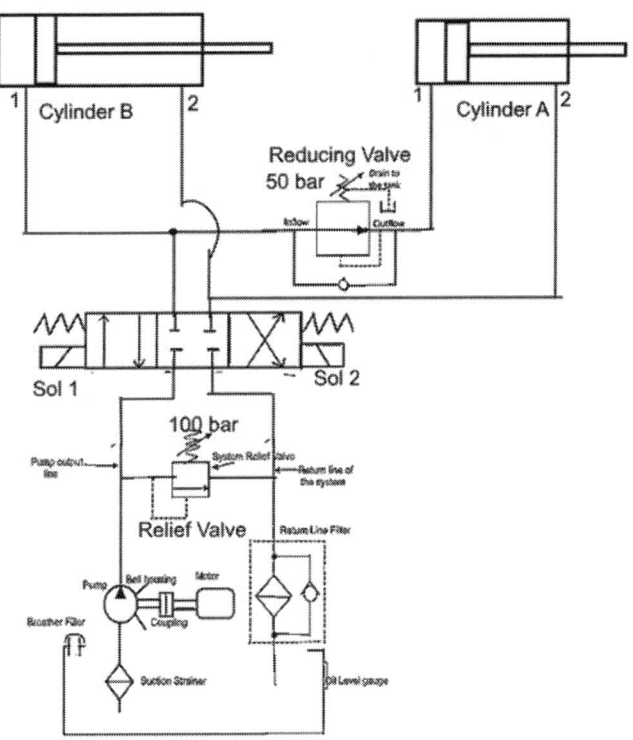

Figure 13 2 Hydraulic circuit with Pressure relef and pressure reducing valve

In the above circuit, let us say that the pressure relief valve is set for 100 Bar. The Pressure reducing valve is to ensure that the cylinder A, clamps the job at a pressure not exceeding 50 Bar.

If the clamping pressure exceeds 50 Bar, the job would get crushed. Let us assume that cylinder B, then drills on the job at a higher working pressure (say) 80 Bar. To shave material of the job. First, let us say Solenoid(Sol) 1 gets energized, after starting the pump. The straight arrow passage gets connected and both cylinders extend. The oil for cylinder A goes through the normally open passage of Pressure reducing valve. It has a sensing element (dotted line) that senses the pressure of the line going to port 1 of cylinder A. The pressure starts to build up on both full-bore areas of the cylinder. In cylinder A port 1 line if the pressure increases beyond 50 Bar, the sensing line lifts the open passage and makes it closed(non-connecting) passage between inflow and out flow. This means that the branch line for cylinder a is sealed at 50 Bar pressure. The cylinder B line has no valve on its line connecting to cylinder ports and the pressure can increase to its operating pressure to complete its shaving the job at 80 Bar pressure.

Since, the job work is completed, the Sol 2 is energized. Both Cylinders A and B get connected through the cross ports of direction control valve. The oil from port 1 of cylinder A will go through the check valve lifting (the ball) and go on to the tank line. Cylinder B, port 1 line also gets connected to the tank line. The purpose of using the pressure reducing valve for reducing the branch circuit pressure of cylinder A is achieved.

13.3 PRESSURE SEQUENCE VALVE IN A HYDRAULIC CIRCUIT

We will consider a similar circuit for explaining the application of Pressure sequence valve. Let us first compare pressure sequence valve with Pressure relief valve.

Both the Pressure relief valve and the Pressure sequence valve have sensing lines from inflow (Pump/pressure lines).
Both are normally closed.
Pressure sequence valve is externally drained and Pressure relief valve is internally drained.
Pressure sequence valve has provision for a return flow through the check valve but not pressure relief valve.

Figure 13 3 Pressure relief valve Vs Pressure sequence valve

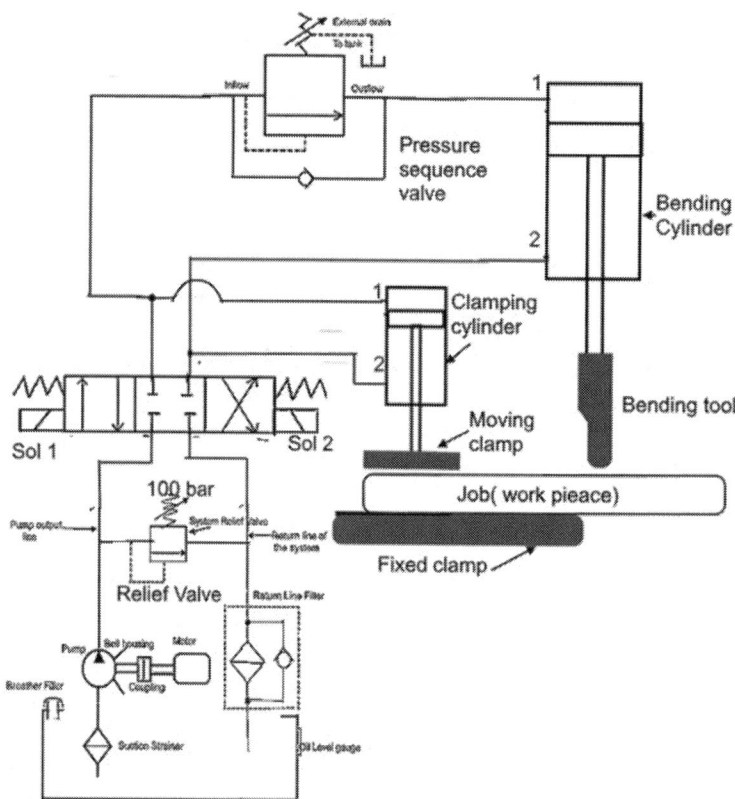

Figure 13 4 Hydraulic circuit with Pressure sequence valve

Please refer above circuit Figure 13 4. In this circuit, we have two cylinders- one for clamping a job and the other for bending the same. The movement sequence to follow is that first the job to be clamped by the clamping cylinder and then the bending cylinder should come down and bend the job. That is the sequence to be followed.

Let us say the clamping is done at 50 Bar and the bending is at 80 Bar.

Once the pump is on, Solenoid 1 is on. The oil is on the pipe lines goes towards ports1 of both cylinders through the straight ports of the direction control valve.

But the oil cannot flow to bending cylinder port 1as the Pressure sequence valve is not open. Therefore, the clamping cylinder gets the oil flow first and starts extending and clamps the job.

As the pressure keeps building up, the same is sensed by the port 1 line of the bending cylinder that has the Pressure sequence valve. This valve is set for (say) 50 Bar and once the line pressure reaches this set pressure, the sensing line lifts the passage to connect the line and the pressurized oil gets in to port 1 of the bending cylinder. The bending cylinder moves forward and the tool bends the job.

Once the job is done, Solenoid 2 is energized and the pump supply gets connected to the cross ports. Port 2 of both cylinders get connected.

Clamping cylinder retracts and the oil from port 1 side of the cylinder goes to tank. In the case of bending cylinder, the oil from port 1 goes to the tank line after lifting the check(ball) of the pressure reducing valve.

Please note that it is possible to use Pressure reducing valve on the clamping cylinder port1 line and use Pressure sequence valve on port 1 line of bending cylinder.

13.4 HYDRAULIC MOTORS IN CIRCUITS

We have not shown the connection of hydraulic motors in hydraulic circuits. In the circuits that follow we shall not show the connection down below comprising of the reservoir, accessories, pump, motor and the Pressure relief valve. It is standard up to Pressure relief valve connection and the student can visualize it, though it will not be shown here to avoid repetition of the same details again and again.

Let us take a hydraulic motor.

Connecting a hydraulic motor would require following additional valves.

1. Over center valve
2. Cross port relief valves

Before demonstrating the hydraulic motor circuit with these valves, let us look at a simple motor circuit without these valves.

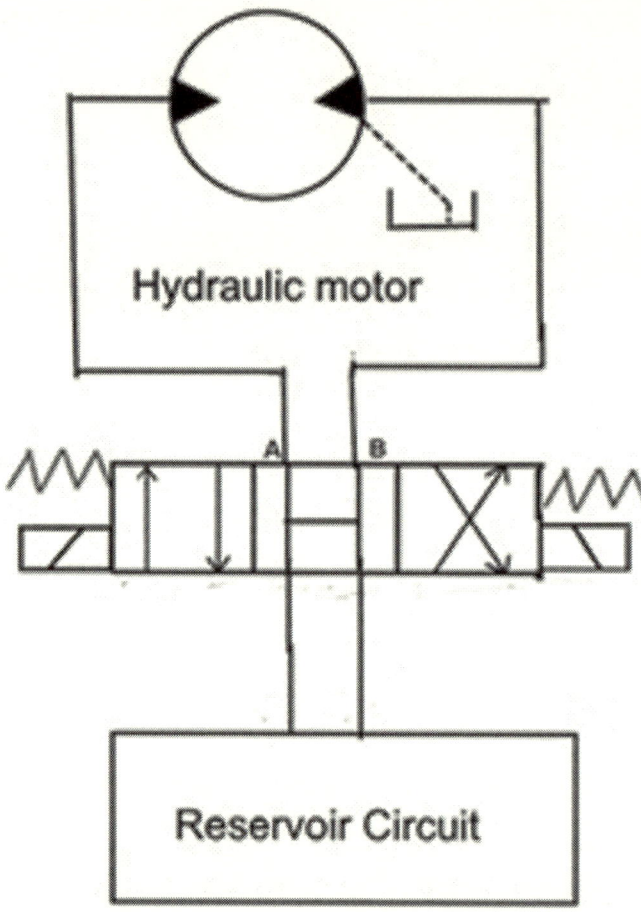

Figure 13 5 A simple hydraulic motor circuit.

13.4.1 Over center valve with a hydraulic motor

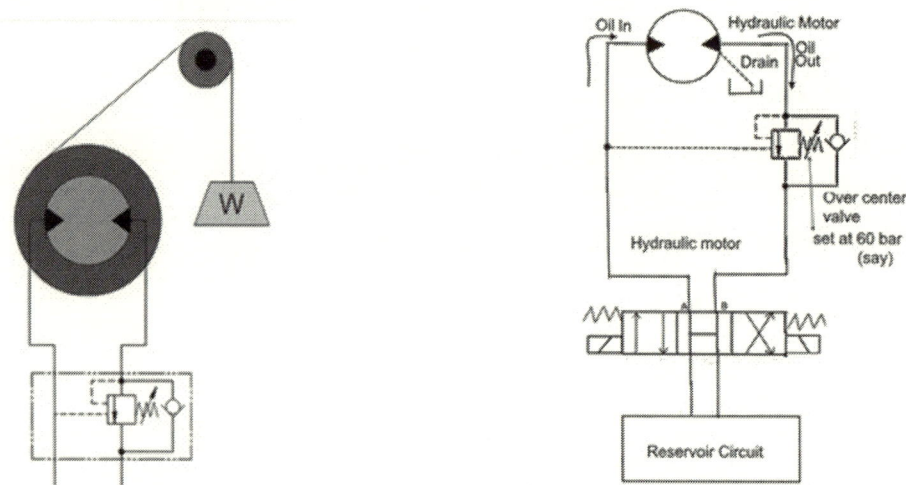

http://www.designworldonline.com/ Figure 13 6 Over center valve application overcenter-valves-are-key-to-hydraulic-control/#_

Over center valve is used when hydraulic motor is used to lower a load(winch operation). The load may act in the same direction as the hydraulic motor direction of rotation. This can result in the hydraulic motor shaft while in rotation would tend to run with (rotate faster)the load. This could mean that the resistance of load is less (meaning the pressure reduction at the input of the hydraulic motor) and the hydraulic motor starts to overrun. To take care of this issue, a over center valve is provided as shown in Figure 13 6.

As explained when the pressure of oil in the out path also becomes less, there is no connection passage for the oil to flow to the tank as the over center valve has a normally closed passage. As a result, oil pressure starts building up before the input to the over center valve. There is a sensing line tapped from the input line to the motor. As per the setting of over center valve, this set pressure would lift the passage to make the connection after the set pressure is reached.

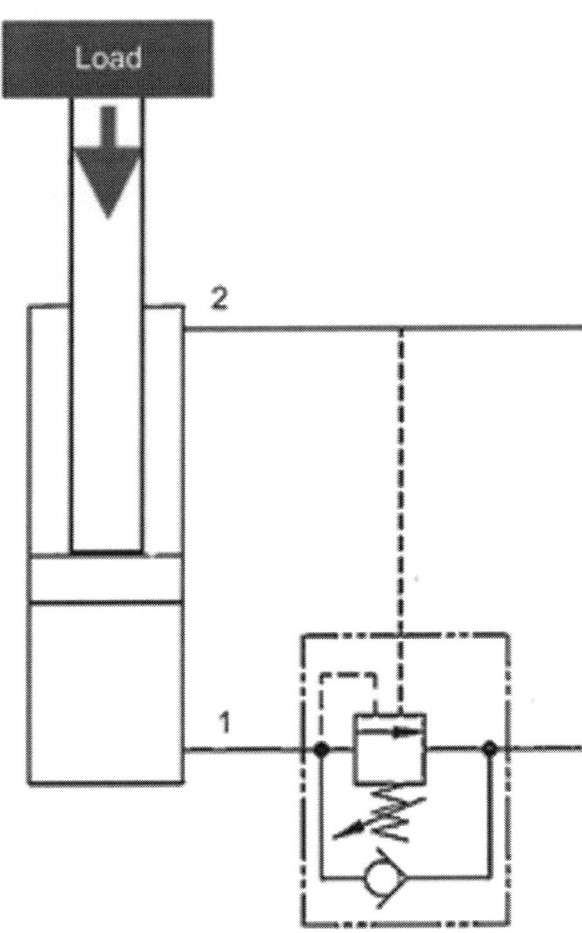

Figure 13 7 Over center valve with hydraulic cylinder

Figure 13 7 shows connection of over center valve in a double acting cylinder return stroke with load has to be managed (situation similar to hydraulic motor running with the load). Talking about free falling load in the case of hydraulic motor or hydraulic cylinder, there is one more valve that is commonly used for this application – it is' Counter balance valve. Most people call over center valve as a counter balance valve.

13.5 HYDRAULIC REGENERATIVE CIRCUIT

A regenerative circuit can double the extension speed of a single-rod cylinder without using a larger pump. With a smaller pump, motor, and tank we can produce the desired cycle time. It also means that the circuit costs less to operate over the life of the machine.

A regenerative circuit can also replace a double rod-end cylinder in some circuits.

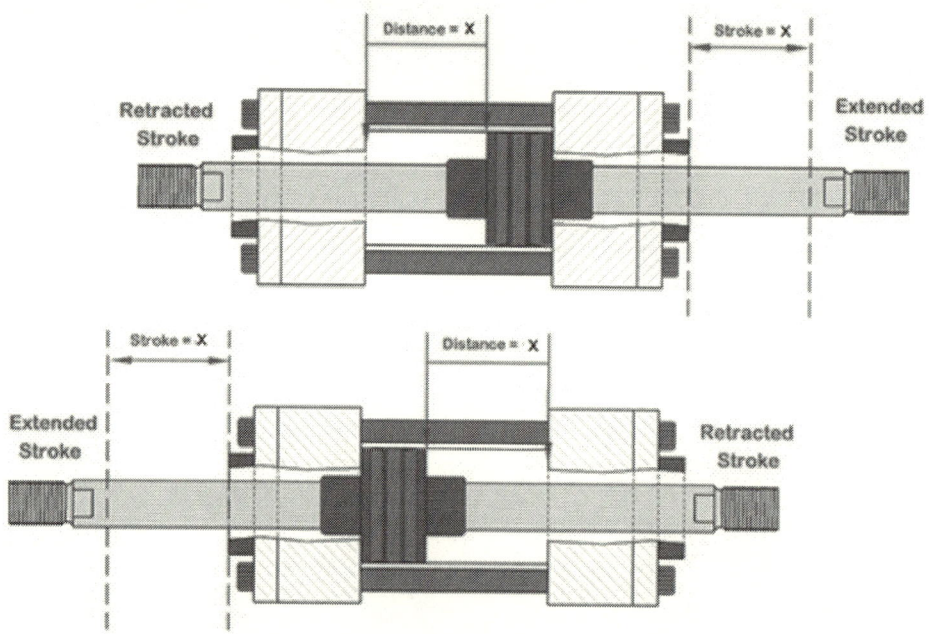

Figure 13 8 Double rod end cylinder

With equal rod diameters, a double-rod cylinder's area is the same on both ends. Equal areas mean identical force and speed both ways at a given pressure and flow. Reciprocating tables often use double rod-end cylinders for this reason. When the main function of a double rod-end cylinder is equal speed and power in both directions of travel, it can be replaced with a regenerative circuit. One requirement is that if a double ended cylinder is to be replaced with a single rod cylinder, then its bore/rod ratio has to be in 2:1 to get more precise equal speed in both directions- of extension and retraction. This means that rod area is half that of the piston area.

With a 2:1 rod, force and speed on the extension and retraction stroke is identical.

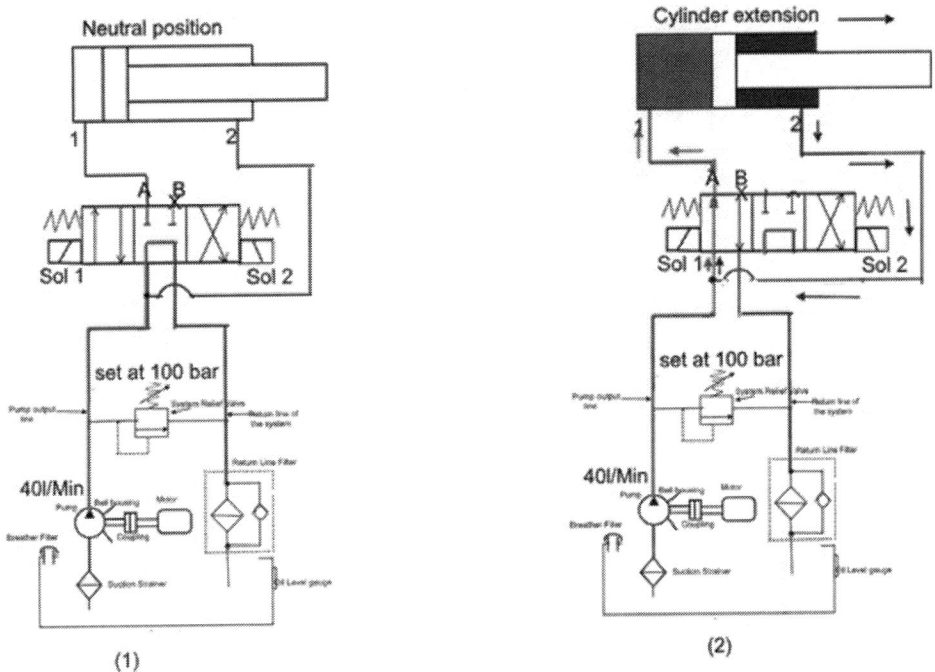

Figure 13 9 Regenerative circuit

Please refer Figure 13 9. The circuit here is explained in the initial position(when the direction control valve is in neutral position) and then, when, the sol 1 is energized. In the neutral position of the direction control valve, P & T lines are connected. Therefore, oil from the pump is getting unloaded to the tank line and is not entering port 1 of the cylinder. The piston is in its home position. Please note that port B of the direction control valve is blocked. This would mean, that, whenever sol 1 or sol 2 is energized, only one port will be working and the second port towards the cylinder will be a blocked port.

Sol 1 is energized.(ref 2) The pump line gets connected to straight passage position of the direction control valve and the oil enters port 1 of the cylinder. The piston/ piston rod starts extending. The oil from port 2 gets connected to the pump line as shown. This would increase the flow rate of oil entering port 1 of the cylinder and the cylinder speed increases because of this additional oil flow from port 2 of cylinder joining the pump flow.

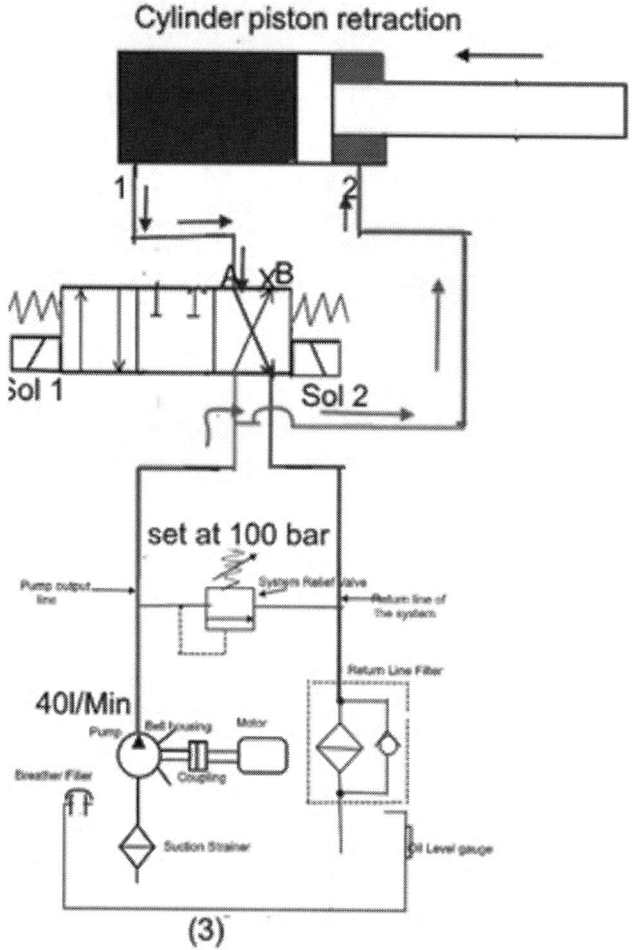

Figure 13 10 Regenerative circuit - piston /piston rod retraction

Please refer Figure 13 10. This figure (3) shows the retraction of the piston rod. It shows the cylinder retracting. Pump flow goes directly to the rod-end port and the cylinder retracts. Oil from the cap-end port returns to tank through the directional valve.

13.6 POINTS TO RECALL

1. understanding the different functions/ symbols of pressure reducing valve , pressure sequence valve and comparison with pressure relief valves.
2. Functions of over center valve and cross port relief valves.
3. Regenerative circuits.

14. FLOW CONTROL VALVES

14.1 LEARNING OUTCOMES

By the end of this chapter, the student will be able to,

1. Understand the function and types of flow control valves.

2. Impact of pressure and temperature on flow of oil and their controls.

14.2 FUNCTION OF FLOW CONTROL VALVES

The purpose of flow control in a hydraulic system is to regulate speed. The flow control valve helps us to adjust the speed of actuators, but, it causes a pressure drop as the flow of oil passes through the valve. Typical application includes regulating cutting tool speeds, spindle speeds, surface grinder speeds etc. Flow-control valves also allow one fixed displacement pump to supply two or more branch circuits fluid at different flow rates on a predetermined basis. basis.

The flow control has impact on the speed of the actuator as well as the power available to the circuit branch where flow control valve is installed.

Speed/ velocity of the actuator-

Q = Flow rate(cubic cm per minute) = a(area in cm2) * velocity-[speed of the actuator(cm/min)]

Power in KW = [Pressure(Bar) * Flow(l/min)]/600

14.3 TYPES OF FLOW CONTROL VALVES

There are two types of flow control valves.

1. Non-pressure compensated flow control valves.

2. Pressure compensated flow control valves.

The non-pressure compensated flow control valve is nothing but a throttle valve or a needle valve where the orifice size can be controlled by an adjusting lever.

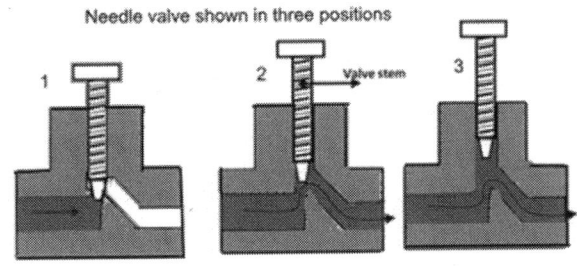

Figure 14 1 Controlling the orifice size by a needle valve

Non-pressure-compensated flow-control valves are used when the system pressure generally remains same and (if used) for hydraulic motor speed that is not very critical.

The disadvantage of valve is that it cannot be effectively used where there is variation in load.

The symbols of non-pressure compensated and pressure and temperature compensated valves are given in Figure 14 2

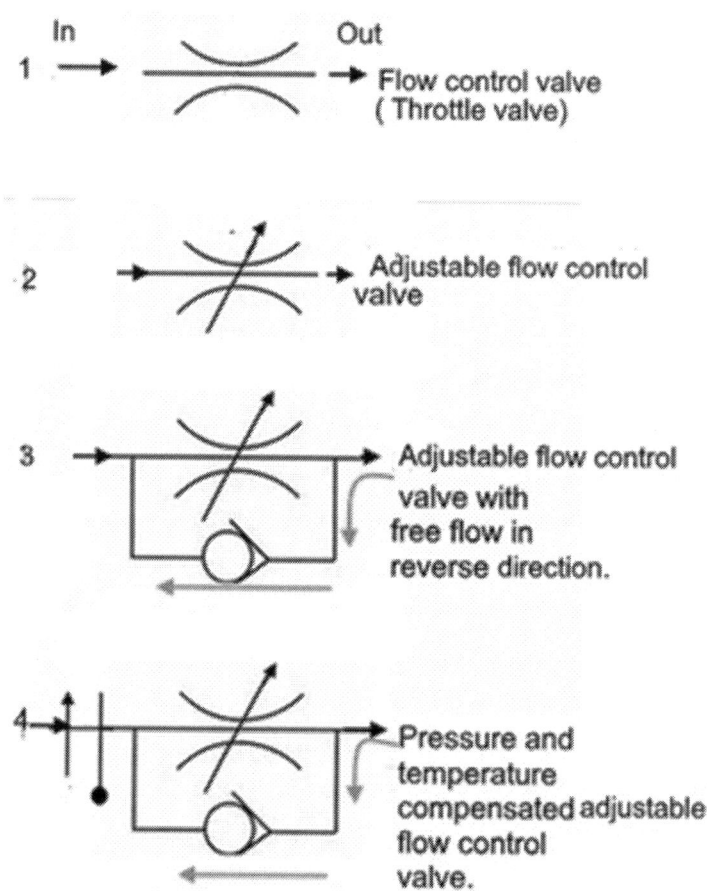

Figure 14 2 Symbols of flow control valve

Figure 14 2 shows that the flow control valves fall into 4 types. The simplest is an orifice, that is not adjustable. In 2, we have an orifice that can be adjusted by a stem attached to a knob. (Figure 14 2) Figures 3 and 4 indicate that we also have pressure and temperature compensated flow control valves.

To explain these types further, we must understand that the flow through the valve is influenced by two factors- Pressure and temperature.

Pressure- The flow through the valve is dependent on the pressure drop across the valve. What causes this pressure differential ?

It is caused by the changes in the orifice size and other obstruction to the flow because of the design of the valve. Greater the pressure differential the greater is the flow. In addition, the load also induces pressure in the line and if the load is variable then the pressure induced by this load also changes and this factor also would cause changes to the flow through the valve.

Temperature- Heat is generated as the oil is used in the hydraulic system. As the oil temperature increases, the viscosity of the oil becomes thin (becomes less viscous) and the flow rate increases. Therefore, the pressure and temperature factors affect the flow control valve and we will not be able to precisely control the adjustment and hence the speed of the actuator with a simple throttle / needle valve. The option to ensure absence in flow variation is to go in for pressure and temperature compensated flow control valve.

14.3.1 Pressure and temperature compensated flow control valve

It is possible to have a pressure and temperature compensated flow control valve as shown by a simplified diagram in Figure 14 3.

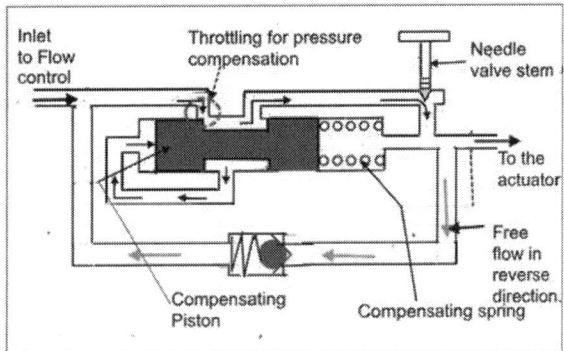

Figure 14 3 Pressure compensated flow control valve arrangement

Pressurized oil enters at the inlet of flow control valve as shown. It takes the path as shown by arrows. The adjustment of flow control is by the needle valve shown.

We have talked about two possible effects of pressure- Pressure drop across the valve and the load induced pressure. Let us say that the pressure drop across the valve is 5 bar. The pressure compensating spring chosen should be of 5 bar rating. The pressure compensating spring moves to push the pressure compensating piston to allow more oil if necessary. For load, induced pressure again, the pressure compensating piston is moved to the left or right for compensating the flow. The figure also shows, that as the cylinder is moving in the opposite direction, then the oil from the port takes the alternative path – it pushes the check valve away and has a free flow path as indicated by the orange arrows.

As far as temperature compensation is concerned t is done by choosing an appropriate metal for the stem of the needle valve. The stem would enlarge and close the valve more when the oil temperature goes up. This means that when the oil becomes thin, (due to temperature), the stem metal would enlarge and close the orifice more. If the temperature falls, the stem metal would contract and allow more oil to flow – thus compensating for temperature variations.

Anti-jump feature in flow control valves: What would happen if the load is suddenly removed. Like in drilling, when the hole opens up- This could cause a sudden jumping of the pressure compensator spool and hence the piston. To avoid this sudden 'jumping 'a stroke adjuster is provided at the pressure compensator end which would limit the sudden movement -This provision is referred as 'anti jump' feature of the flow control valve.

The discussion so far is using the flow control valve when a fixed displacement pump is installed in the system. If there is an option to use a variable delivery pump, then the flow rate gets adjusted due to the load induced pressure variation by the pressure compensation provision in the variable delivery pump itself. There is no need to go in for a separate flow control valve for the system. However, variable delivery pumps are more expensive and it would be a cheaper option to go in for a fixed deliver pump ad the flow control valve. The choice depends on the application.

14.4 POINTS TO RECALL

1. Constructional arrangements of pressure and temperature compensation in flow control valves.
2. Need for temperature and pressure compensation in flow controls.

15. UNDERSTANDING HYDRAULIC CIRCUITS-4

15.1 LEARNING OUTCOMES

By the end of this chapter the student will understand,

1. Different methods of flow control in a hydraulic circuit using flow control valves

15.2 FLOW CONTROL VALVE USING METER IN METHOD.

We have already seen the symbol of flow control valve in Figure 14 2. While fixing the flow control valve in a circuit following points to be considered.

1. The flow control valve is generally fixed at a location closest to the actuator- that is, between the direction control valve and the actuator.

2. The check valve position has to be appropriate for the application(meter in or meter out)

Please refer Figure 15 1

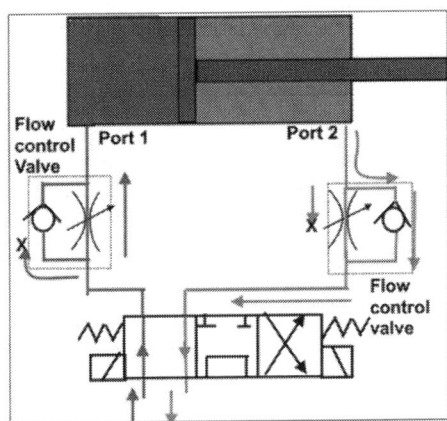

Figure 15 1 Meter in control circuit

The most important point is while connecting the check valve position for meter in application, the check valve ball should not allow the oil to pass through it.

In Figure 15 1, The circuit portion above the reservoir circuit is only shown. Pressurized oil from the pump (of the reservoir circuit) goes through the straight ports portion of the direction control valve –as indicated by red arrows.

At the entry of the flow control valve, the oil has two path options. It can either go through the narrow-restricted path way straight ahead (shown wit hred arrow head) or through the check valve. But the check valve route is not possible as the ball is blocking its path.(shown by X letter)

Then the only option is through the restricted, controlled(adjustable) narrow path. This would mean that less oil gets into the full-bore area of the cylinder. Because less oil gets in, the speed of the cylinder will also be less than the normal speed(if full oil without this restriction is allowed).

The speed is adjustable by adjusting the narrowness of the path of the flow control valve. The black arrow indicates the adjutability of the narrow path of the flow control valve On the other side of the actuator,(green side indicating the oil to be exhausted back to the reservoir) the oil also has to go through the flow control valve. It can either go ahead through the restriction or through the check valve. Here, the check valve cap is on the other side of the ball. This would mean that the oil can easily shift the ball away and freely flow out.

As per properties of the fluid, the oil always chooses the path of least resistance, the oil flow will be through the check valve portion instead of the restricted path portion. In effect this means that the oil flow restriction is only in the inflow to the actuator and not during the outflow from the actuator. Hence the circuit is referred as 'Meter in circuit. The meter out circuit is the same but the flow control valve orientation interms of check valve has to be changed.

To make the understanding even simpler, please look at Figure 15-2.

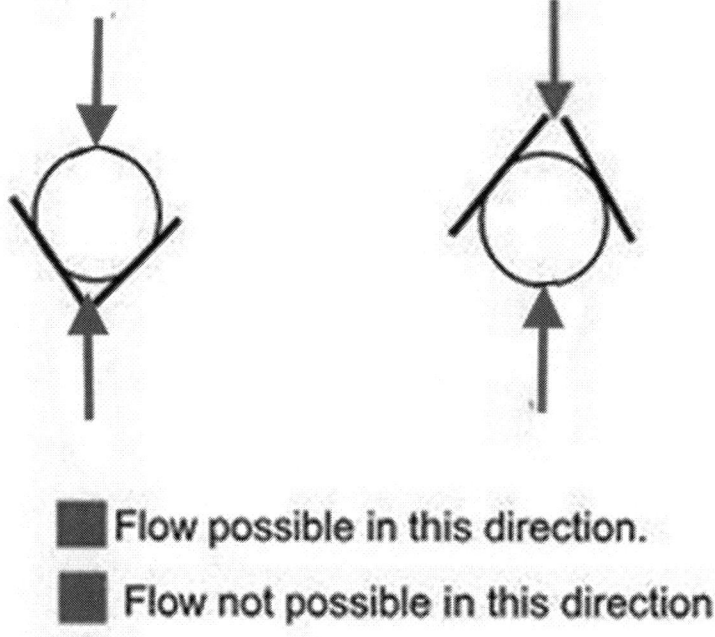

Figure 15 2 Check valve orientation in flow control valve.

Depending on the orientaion of the cap of the check valve and the direction of flow, the check valve allows the flow or does not allow the flow.

15.3 METER OUT CIRCUIT

Meter out circuit is controlling the oil outflow from the actuator.

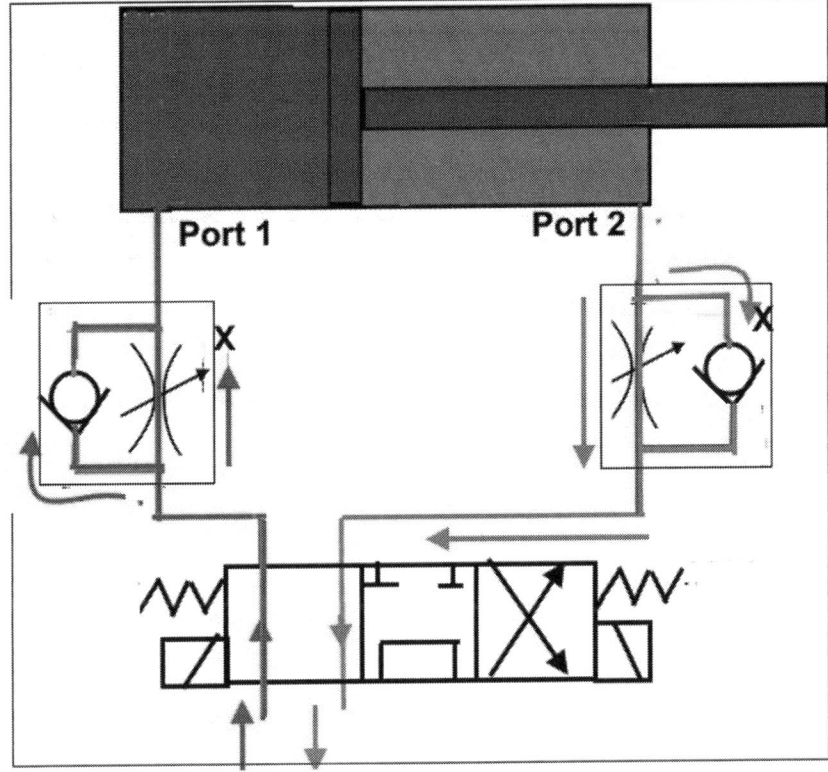

Figure 15 3 meter out circuit

Compare Figure 15 1with Figure 15 3. You would notice the only difference is in the orientation of the check valves. In meter in circuit the caps on the valves in both inlet and outlet is on the head of the ball. In the meter out circuit the caps(the cross hair lines) are on the bottom pf the balls.

What that means is that in the meter in the meter out circuit, the inflow to port 1 is free as the flow lifts the ball of the flow control valve and is on its way. While the oil that should go to the tank has to be metered –controlled. (Tthe oil is not freely allowed to be drained through the check valve portion of flow control valve), this offers resistance to the movement of the piston and hence the speed is restricted/controlled.

Another important aspect is that in meter in and meter out circuit it is not necessary to have flow control valves on either side off the actuator. They are drawn as shown to increase the understanding with respect to the orientation of the check valve.

In the meter in circuit, it is enough the flow control valve is drawn only on the port 1 line. For meter out circuit the flow control valve as shown in Figure 15 3, is drawn on port 2 line of the circuit.

15.4 BLEED OFF CIRCUITS

Please refer Figure 15 4.

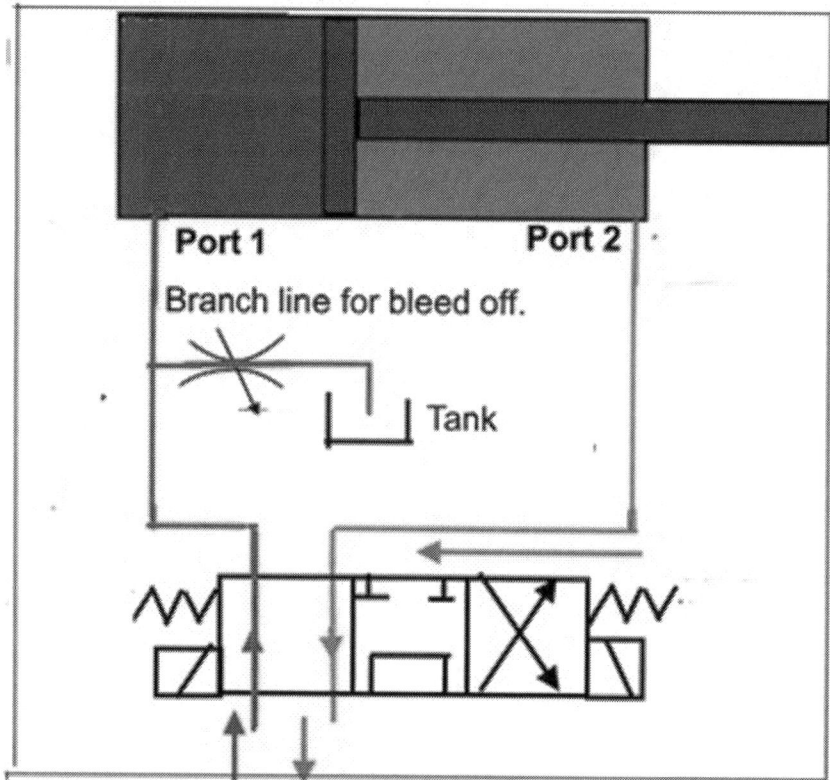

Figure 15 4 Bleed off circuit

The purpose of bleed off circuit is the same as the meter in/meter out circuits.. To control the speed of the actuator. In the Figure 15 4, on the inlet port 1 line, an extra branch line for the oil is provided.

As the oil has an alternative path, part of the oil gets drained to the tank. Only part of the oil goes inside port1 and hence the speed of the actuator is limited. The quantity of oil that gets drained to the tank can be adjusted by the flow restrictor valve. Please note that in the symbol, no check valve is shown in the flow control valve. However, if the flow control valve with check valve is only is available, then the check valve orientation has to be correct. The flow should only go through the restrictor and not to be through check valve.

We have so far seen speed control using flow control valves. In special purpose and other machines, we sometimes have to think of a tool post carrying a tool to approach the job (work piece) fast and then have to do the feed (cutting/drilling/milling etc.,) operation at a lower speed.

These feed circuits can be of different types as follows.

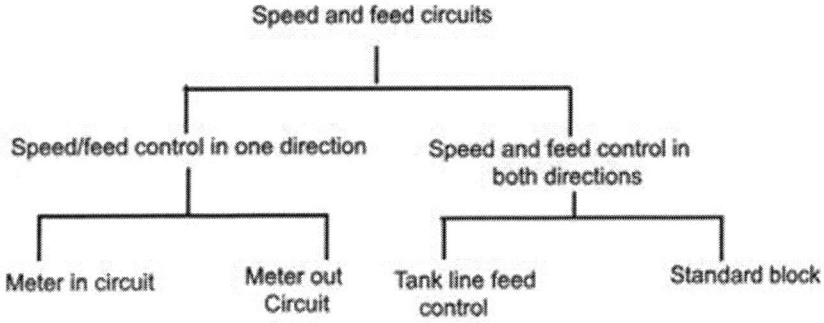

Table 15 1 Types of feed control circuits

15.5 SPEED AND FEED CONTROL IN ONE DIRECTION- MACHINE TOOL APPLICATIONS

The application involves the tool making a fast approach then feed at a reduced level of movement rate and then fast return.

In the flow control application circuits, we have seen earlier, the flow control restricts the speed at a particular rate of movement. We do not have a dual speed movement in one direction of the actuator. In machine tool applications, we would need movements in such a way that, for example, the tool post should move at a faster speed and once at the job, the movement has to be at a much lower speed to carry out the desired operation of cutting or drilling etc., (feed speed). Thus there is a fast approach speed and a feed speed. The requirement could be in one direction of the actuator or in both directions of extension and retraction.

In the next two circuits that follow we shall see the dual speed approach, first during the extension of the piston rod(meter in) and then during the retraction movement(meter out).

15.5.1 Speed and feed control – Meter in circuit

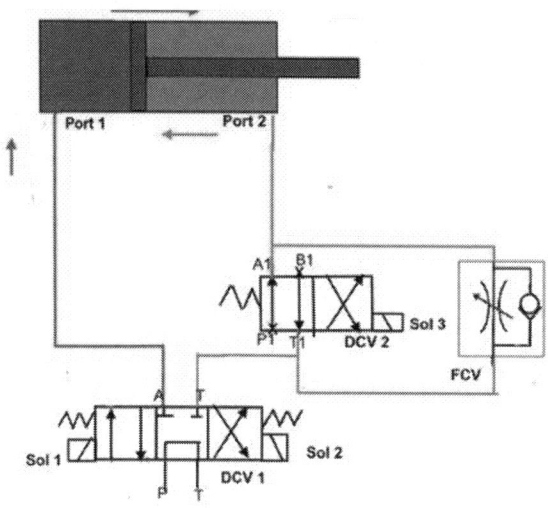

Figure 15 5 Dual speed control - Meter in

Please refer Figure 15 5. The circuit is shown from the main 3 position direction control valve onwards. Below the direction control will be the reservoir circuit.

The cylinder is shown with extension movement as Solenoid 1 is activated- straight arrow side gets connected and the oil goes to FCV restricted passage way path(P1 is blocked and B1 is also blocked in the two position direction control valve (Sol 3 is not activated t the current situation) This would result in slow extension of the piston rod.

Oil from the other side of port 2 has no restriction and gets drained.

When a fast approach is required, then Sol 3 is energized. This would make cross port position of the two-position direction control valve to get connected- T1 will get connected to A1 and then A2, thus by passing the flow control valve. Thus faster speed is achieved.

Thus, we can have both fast approach and feed speed can be obtained by the hydraulic circuit shown above.

15.5.2 Speed and feed control – Meter out circuit

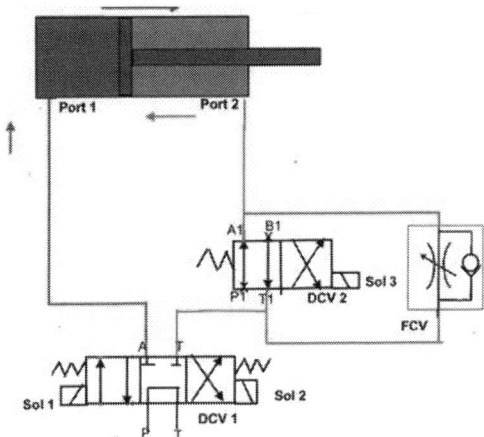

Figure 15 6 Dual speed control - Meter out

Please refer Figure 15 6. As explained in the previous circuit, the connections are shown from the main 3 position direction control valve onwards. The cylinder shown advancing after Sol 1 is engaged. There is no restriction for the oil to flow to port 1 - The cylinder should extend in normal speed if there is no restriction for the oil to flow out through port 2 of the cylinder.

But, here there are two pathways for the oil to flow back to the tank. The one shown by green line indicates the path through the flow control valve -restricted path. As the oil flows through the restriction, the piston/ piston rod is not able to extend freely(though the oil inflow has no restriction – out flow being restricted)- feed speed is obtained. This can be adjusted by the adjustment knob of the flow control valve.

When we need fast speed, we must be in a position to take an alternative path by passing the flow control valve. This is possible by energizing Sol 3. Then, A1 would get connected T1 because of the cross port activated by Sol3. Once this easy path is available, oil flows drains out to tank line avoiding the flow control valve path.

Thus, fast speed is obtained.

15.5.3 Speed and feed control in both directions of piston movement

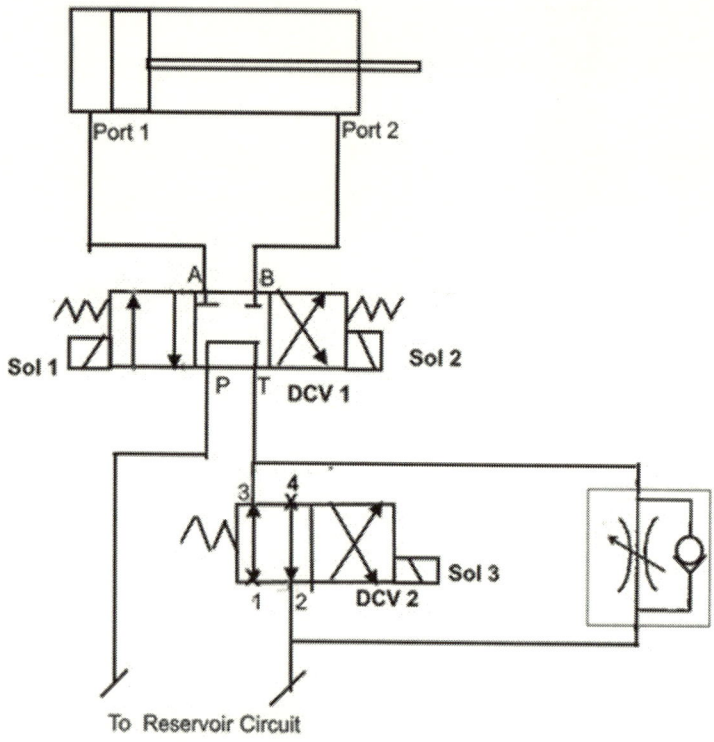

Figure 15 7 Speed and feed control in both actuator directions.

This circuit in Figure 15 7 is similar to the previous circuits. The change is that we have made the two-position direction control valve and the flow control valve path options available to port 1 and port 2 lines as and when the feed speed is required.

If only rapid speed is required, Sol 1(3 position main direction control valve) and Sol 3 are energized. The extension is at a faster speed and the oil flow from port 2 finds the path 3 to 2 through the cross-port passage of two position direction control valve.

If feed speed is required in the same direction, then oil from port 2 goes through the flow control valve (Sol 3 is not energized).

In the same manner if faster retraction is required cross-port (Sol 2) side of the main direction control valve is energized and the oil flow from port 1 can go through cross- port side of two position direction control valve(by energizing Sol 3) and the faster retraction is possible. If slower feed speed is required during the retraction movement, then, the oil flow is made to go through the flow control valve by not energizing the Sol 3.By this hydraulic circuit, it is possible to have both rapid approach and the feed speed during the extension and retraction stroke of the cylinder.

15.5.4 Standard manifold for dual speed

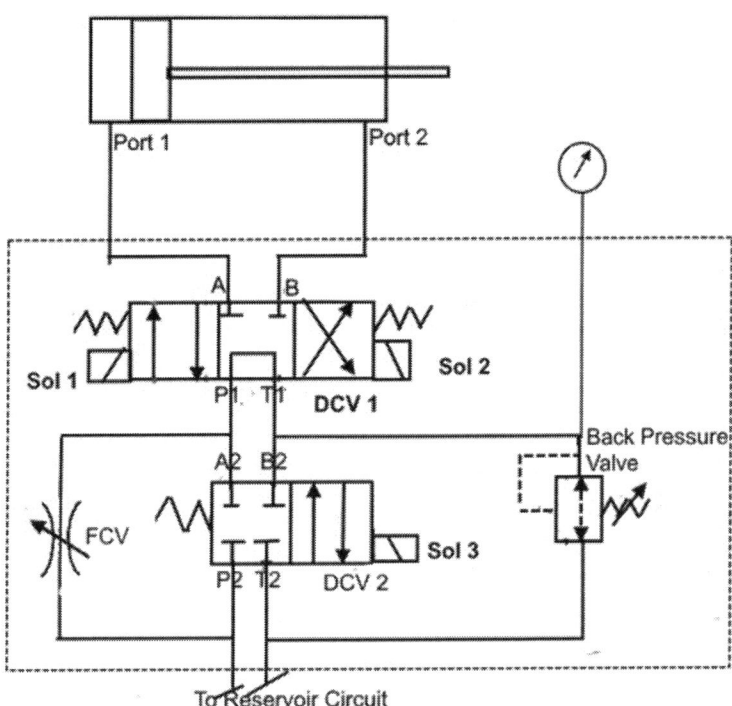

Figure 15 8 manifold for dual speed

The standard manifold method is perhaps one of the options manufacturing machine tools with regular requirements of speed/feed systems in both direction of the actuators.

Looking at the circuit in Figure 15 8, it should be clear how the system works with both rapid speed and feed speed.

During the cylinder's forward stroke, for rapid(normal- unrestricted) speed, Sol3 and Sol1 are energized. This would facilitate a clear path of the pumped oil to go to port 1 and the extension is achieved. If during part of the extension, feed speed is required, then, at the set time, Sol 3 is released (cut off) and the oil has P2 port blocked and hence is routed through the flow control valve – less oil is flowing to port 1 through the straight passage position of 3 position, main direction control valve – feed speed is obtained.

During the return stroke, for normal speed Sol 3 and Sol 2 are energized. For feed speed, sol 3 is cut off for administering less oil through port 2 and thereby feed speed is achieved.

15.6 HYDRAULIC CIRCUIT FOR PRESS APPLICATION USING DOUBLE PUMP

We have seen, in the earlier circuits, getting dual speed by the arrangement of valves. It is possible to achieve the same result by installing a double pump. A higher capacity pump at low pressure would make the cylinder move at a rapid speed. – Then, we can cut off this high flow rate pump and activate a high pressure, low capacity pump to make the cylinder deliver the feed portion of the work cycle.

Let us assume that the hydraulic circuit designer has the following data to design a hydraulic circuit for a baling press. (Baling - to compress large quantities of cotton or waste material and make them smaller).

1. Double acting hydraulic cylinder is to be used.
2. Pressing force should be 100 tons.
3. Rapid speed must be 2.2 meters / min.
4. Pressing speed to be 0.3 meters / min.
5. Solenoid operated direction control valve to be used.

The hydraulic designer must also get as much additional information about the press like day light of the press (The height/ distance between the fixed and moving plates at the fully retracted position of the cylinder, the plate weight, duty cycle of the machine etc.,

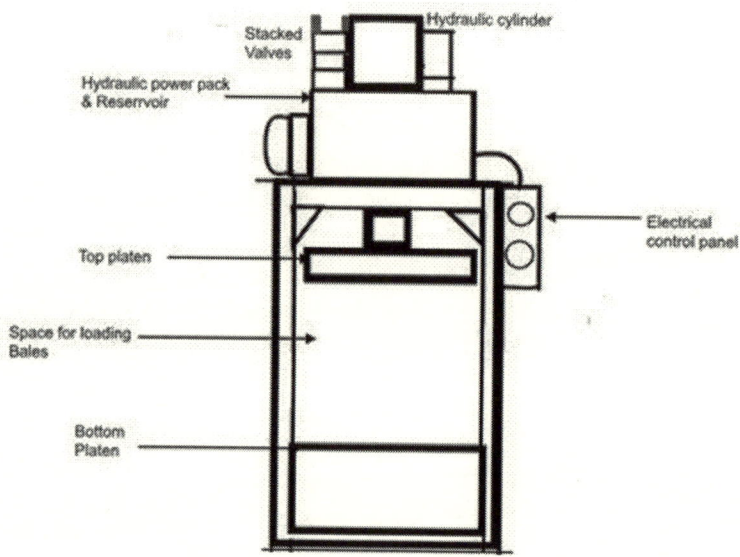

Figure 15 9 Baling press structure

Calculating cylinder specifications-

Working pressure = Load/ full bore area of the cylinder

Let us assume the working pressure is 250 Kg/Sq.cm

250 Kg/cm2 = 100 tons/ area of the cylinder = 100,000 Kg/ Cylinder full bore area

Cylinder full bore area = 100,000/250 = 400 cm2

400 cm2 = $[\pi D^2]/4$ where D is the full-bore Diameter of the cylinder.

D = 22.57 cm = 225 mm

The cylinder manufacturers have standard bore and rod diameters of the cylinders.

The closest cylinder bore suitable from manufacturers range (say) = 250 mm

Standard rod size for above = 140 mm.

Calculating actual working pressure = Load / area = $[100{,}000 \text{ Kg}] / [\pi D^2]/4$; D=25cm

= 203.8 Kg/cm2

Flow rate of the pump required (l/Min.)=Q=

= (Full bore area of the cylinder X velocity)/1000

We shall consider fast approach velocity for this calculations as it might require higher capacity pump.

= (490.6 cm2 X 220cm/min) /1000 = 110 l/min.

Let us calculate the pump flow rate required for feed rate given, that is, 0.3 m/min.

= 490.6cm2 X 30cm/min/1000 = 15 l/min.

Calculating horse power required

For a fast approach of 2.2 m/min, we need a pump of 110l/min,(During fast approach, there is no load pressure, and only minimum pressure required to move the piston/ piston rod(The load pressure starts once the job is pressed by the actuator)

Power required in KW = PQ/600(PQ/600 is a thumb rule for calculating power in KW)

= (110 Kg/cm2 X 110l/min)/600 = 1.8 KW ≈ 2.34HP Next higher rated motor available is 3HP

For pressing speed, pressure = 204 Kg/cm2(The actual working pressure is only 203.8kg/cm2) and Q = 15 l/min

= (204 Kg/cm2 X 15l/min)/600 =5.1 KW ≈ 6.63 HP Next higher rated motor available is 10 HP considering a safety factor

In case we go in for a pump that gives 110 l/min and at a pressure of 204 Kg/cm2,then the motor HP requirement for this pump = (204Kg.cm2 X 110 l/min)/600

= 37.4 KW = 48.62 HP ≈ 60 HP, considering a safety factor- This could be an expensive investment compared to a double pump. If we can go in for two pumps(double pump configuration available from manufacturers)-One pump for high volume at low pressure(110 l/min at 10 Kg/cm2) and another for low volume and high pressure(15l/min at 204 Kg/cm2) both driven by one motor , we can save considerably on power consumption. For a double pump operation, we must select a motor that is of higher rating of the two HPs we have calculated that is of 3HP and 10 HP, choose, 10 HP.A 10 HP motor coupled to two pumps (110 l/min at 10 Kg/cm2 and 15l/min at 204Kg/cm2.Can meet the functional requirements of this application.It is essential that our hydraulic circuit is made in such a way that high volume low pressure pump is initially on and once the platen touches the bale of cotton/ waste materials, it starts the pressing speed at higher pressure. As mentioned earlier double pumps are available from manufacturers.

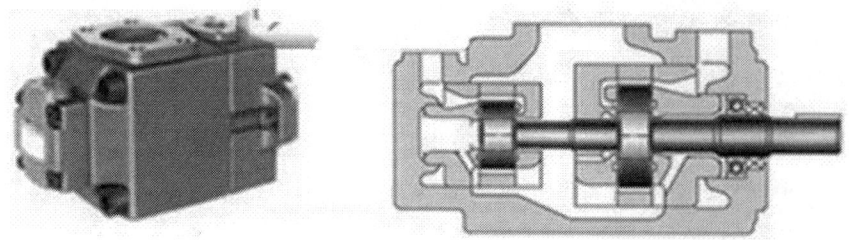

Figure 15 10 Double vane pump - PV2R series of yuken India

The hydraulic circuit for the above application Figure 15 11.

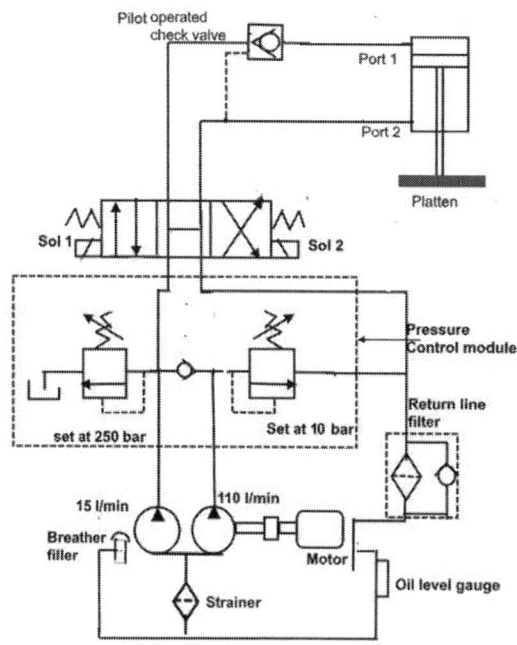

Figure 15 11 Hydraulic press circuit for press application

Here in the circuit diagram of Figure 15 11, we have used a double pump and a pressure control module. The pressure control module is a specially designed attachment that is available to work as a relief valve for the double pump circuit.

It has two relief valves in alignment and is connected between the pumps and the tank as shown above.

In the beginning, the cylinder piston rod is in the initial home position. When the motor is on, both pumps starts delivering the oil and the total delivery will be 125 l/min(15 l/min+ 110 l/min)The bigger pump lifts the check valve and takes the path. Sol 1 is energized and the straight arrow position is activated. Flow of 125 l/min goes to port 1 of the cylinder through the pilot operated check valve, lifting the ball of the check valve. The piston starts moving at a rapid speed and as soon as it touches the bale of cotton/ waste materials there is resistance to movement. Pressure increases and the relief valve set at 10 bar unloads 110 l/min pump flow to the tank.

Only 10l/min pump delivery goes to the cylinder, as its relief valve is set at a much higher pressure of 250 bar- and the baling press get pressed. We have used a relief valve set at 250 bar which is higher than the working pressure. It is possible to set the relief valve at any higher pressure between the working pressure and 250 bar.

The pilot operated check valve in port 1 line ensures that there is a hermitic sealing of the oil in the cylinder. Application of POC requires that we use all ports connected (in neutral position) direction control valve in the circuit.

Retraction of the cylinder is done using Sol 2 of the 3-position direction control valve.

Oil from port 1 is pushed out and the pilot operated check valve allows the flow because of the pressure from the pilot line tapped from port 2 line.

The power requirement and saving in case of double pump can be graphically explained for this application.

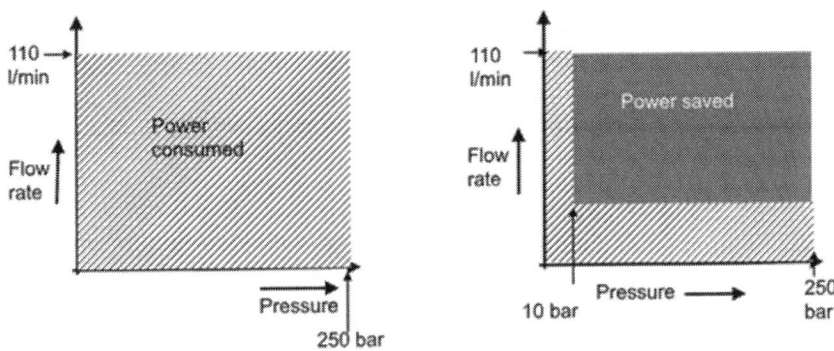

Figure 15 12 Power saving in double pumps

Please refer Figure 15 12, in the first graph, the total power consumption is calculated for 110l/min and the max. pressure of 250 bar- which would work out to 45.3 KW.

When we use a double pump, the power saving is shown. It is mainly because when the flow is 110l/min, the pressure is only 10 bar. When the pressure is at 250 bar the flow is only 15l/min.

Hence double pump is a preferred type of pump when applications like presses are considered.

15.7 POINTS TO RECALL

1. Types of flow control valves, their symbols.
2. Methods of speed control in hydraulic applications.
3. Double pump application and power savings.

16. ACCUMULATOR APPLICATIONS

16.1 LEARNING OUTCOMES

By the end of this chapter, the student will be able to understand

1. The applications of hydraulic Accumulators.

2. Hydraulic Accumulator circuits.

16.2 FUNCTIONS, TYPES AND SYMBOLS OF HYDRAULIC ACCUMULATORS

The function of hydraulic Accumulator is to store pressurized oil in a special container (Compare with the reservoir, where oil is stored in normal conditions. In certain applications, we would use both the reservoir and the Accumulator. Reservoir is in almost in all applications, whereas the Accumulator is used only against certain requirements, when pressurized oil is required for short durations of time.

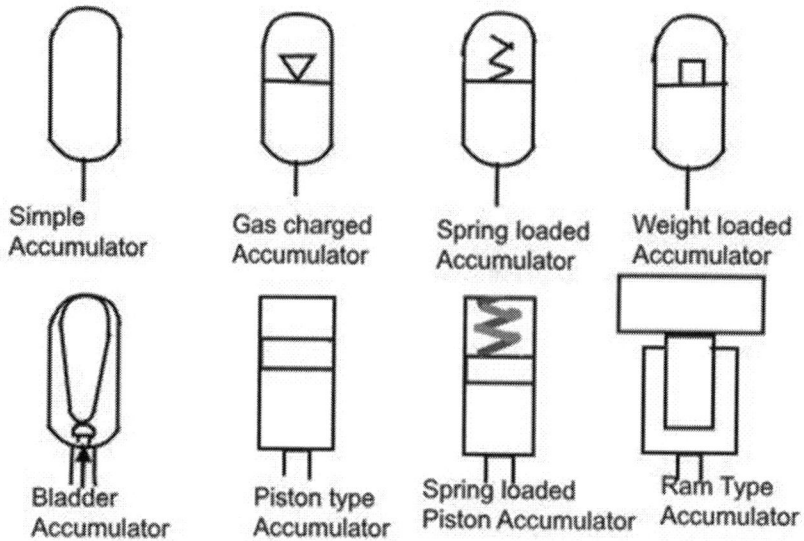

Figure 16 1 Types and symbols of Accumulators

Hydraulic Accumulators are used normally in following circumstances.

Most hydraulic accumulators are used in one of four applications:

1. Supplement pump flow in circuits with medium to long delays between cycles.

2. Hold pressure in a cylinder while the pump is unloading or stopped.

3. Have a ready supply of pressurized fluid in case of power failure.

4. Reduce shock in high velocity flow lines or at the outlet of pulsating piston pumps.

We shall discuss the most popular types of Accumulators(Accumulators using gas and oil are known as Hydro Pneumatic accumulators) used in this chapter.

16.3 BLADDER TYPE ACCUMULATORS

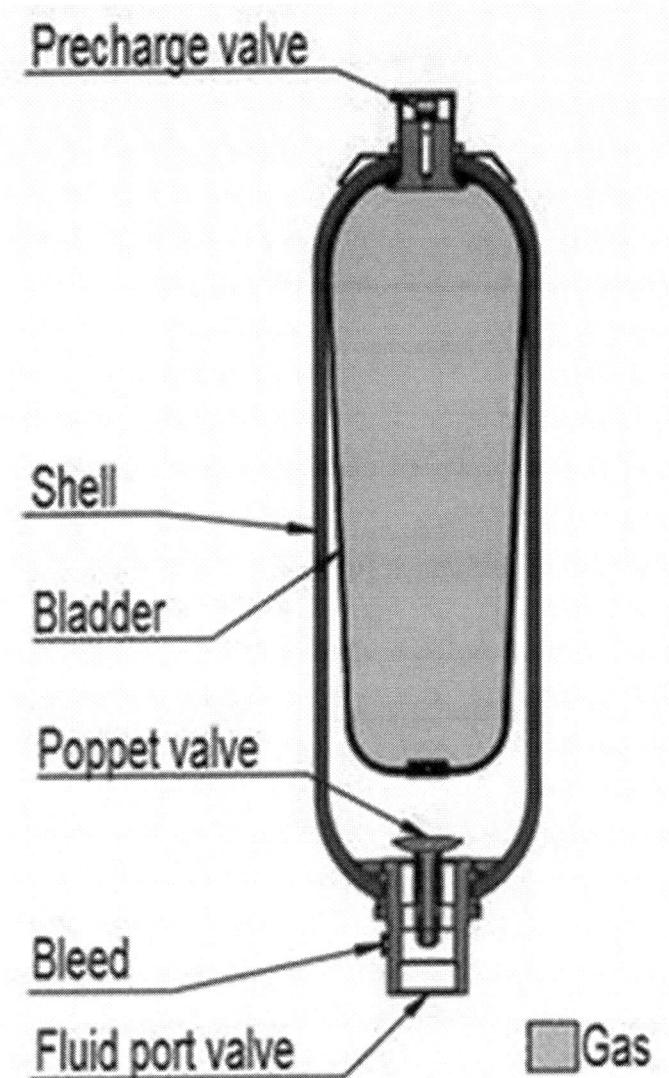

Figure 16 2 Bladder type Accumulator

Please refer Figure 16 2. Here the bladder contains an inert gas(Nitrogen). The figure shows the bag inside a cylindrical shell. The oil is outside the bladder filling the rest of the space in the shell. The gas is filled through the prefill valve. If there is a pressure increase in the system, the pressurized oil enters through the poppet valve shown, and it gets inside compressing the gas in the bag and thereby increasing its own space in the accumulator. If there is a drop-in pressure in the system and the pressurized fluid in the accumulator rushes out (causing expansion of the gas which was earlier compressed.)

The process of filling the gas with pre-charge valve and the oil entering the shell when there is a pressure increase and rushing out when there is a pressure drop is shown in the following figure.

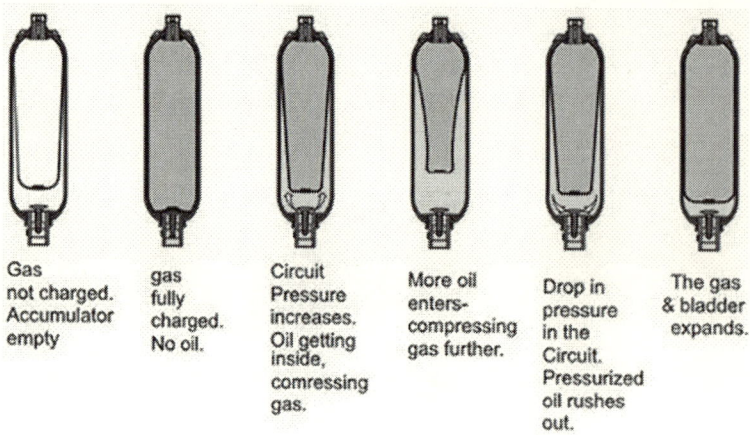

Figure 16 3 Bladder type working process

Piston type Accumulator process of working given below.

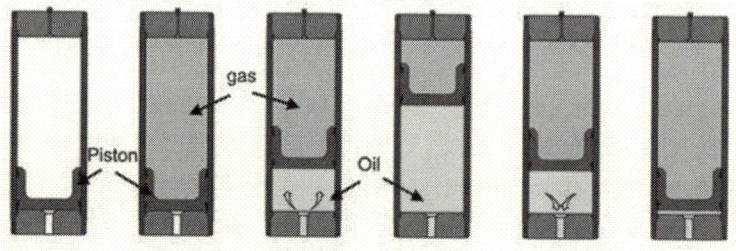

Figure 16 4 Piston type Accumulator working process

Similar to the bladder type Accumulator is the diaphragm type as shown below.

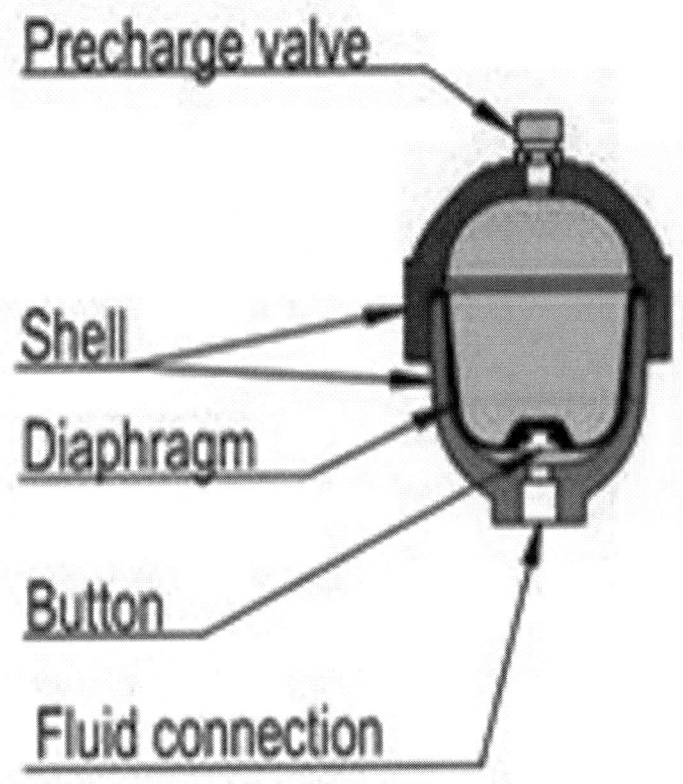

Figure 16 5 Diaphragm type Accumulator

16.4 SIZING -ACCUMULATORS

In respect of bladder/diaphragm type the sizing can be done using the following:

Let $p0$ = Nitrogen gas pre-charged pressure

$p1$ = Minimum working pressure

$p2$ = Maximum working pressure

All above relate to Nitrogen gas pressures.

Normally, $p0 = 0.9\ p1$ and

$p2 \leq 4\ p0$ (for bladder and diaphragm types)

Further, if,

$V0$ = gas volume corresponding to $p0$

$V1$ = gas volume corresponding to $p1$

$V2$ = gas volume corresponding to $p2$

Then, $\Delta v = V1 - V2$

As we are dealing with state of gas changes, for isothermal change,(Assuming the change in the gas inside the accumulator takes place in such a way that there is sufficient time for the complete exchange of heat between Nitrogen and the surroundings)

$p_0 V_0 = p_1 V_1 = p_2 V_2$

For adiabatic change(the change in pressure/ volume takes place rapidly),

$p_0 V_0^n = p_1 V_1^n = p_2 V_2^n$

Where n =4(adiabatic constant for Nitrogen)

Normally the Accumulator manufacturers furnish the curves relating to p_0 p_1 and p_2 and the sizes of the Accumulators. The hydraulic designer should know the values of p_0 p_1 and p_2 based on the requirements and can select an appropriate size of the Accumulator.

16.5 ACCUMULATOR CIRCUITS

The popular Accumulator symbol and its connections to the circuit is used in the circuit is as follows.

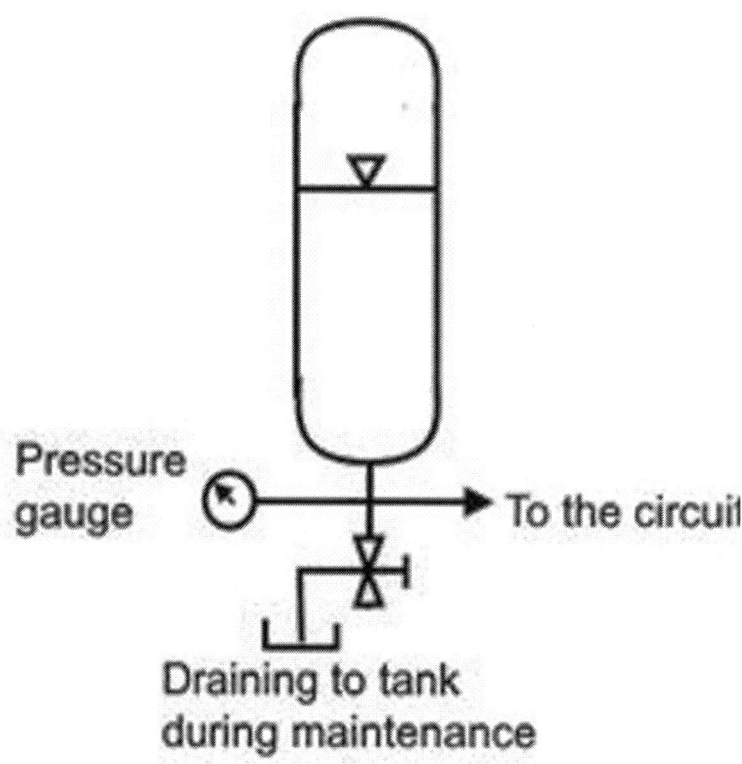

Figure 16 6 Accumulator symbols and connection

16.6 LARGE FLOW RATE REQUIRED FOR SHORT PERIODS – APPLICATION IN PLASTIC INJECTION MACHINES

The graph shows the flow requirements of a plastic injection machine during one cycle. The requirements peak during Q2 and for a short period of t2. For large part of the cycle the flow rate is less at Q1. The average requirement during the cycle is indicated as Q.

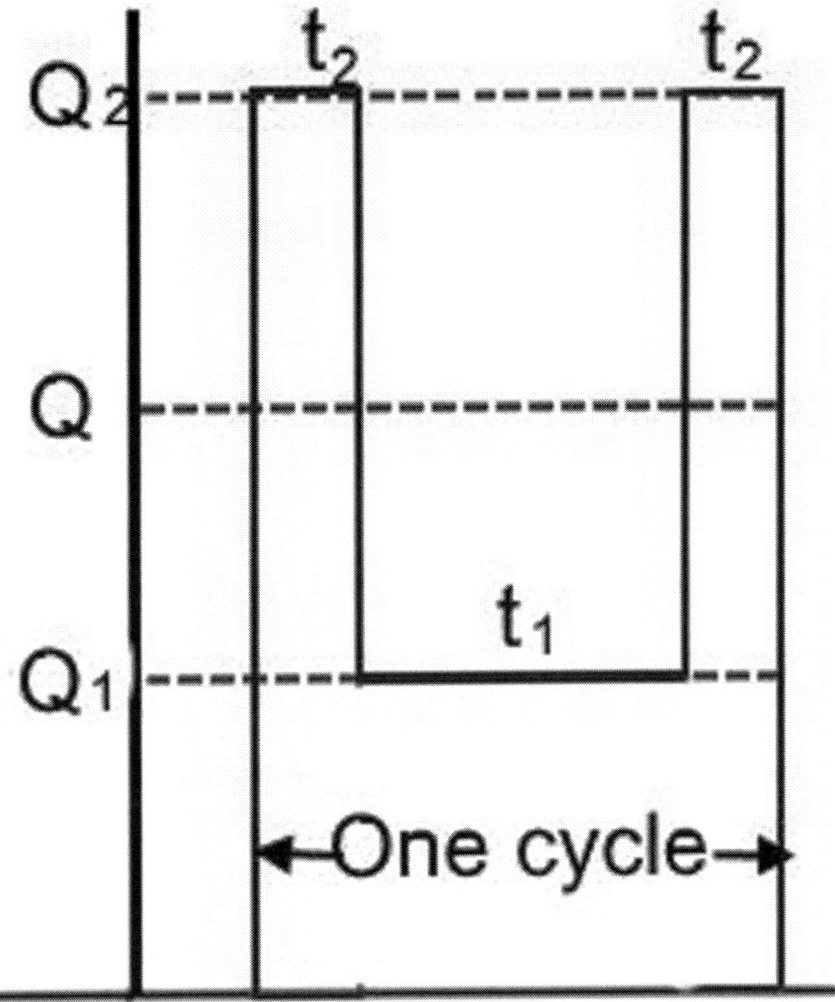

Figure 16 7 Rate of flow during one cycle.

It might be uneconomical to go in for a large sized pump to meet this requirement. Accumulator in the circuit could provide for short bursts of large volumes of water.

The power (and thus the flow rate) of the pump can be instead fixed according to the average absorption. In the early stages of the working cycle when the needs of system flow rate less than the pump, the accumulator is filled. When the maximum flow rate is required, the difference in comparison with the pump supply is taken from the accumulator.

Let us understand the effectiveness of the accumulator for the above application.

Let us assume in the above cycle (Figure 16 7), that the additional flow is required during t_2 is 5 seconds and the passive part of the cycle when no additional oil is required t_1 is 20 seconds. Assume that the Accumulator chosen can supply 15 l/min. If it supplies this fluid during the active part of the cycle, it must be recharged during the passive part (t_1 portion of the cycle).

Therefore the pump capacity for this compensation should be ,

Q = 15 l/min during 20 seconds' cycle.

Q = [15 l/min] ÷ 20/60 = 45 l/min. (should be the capacity of the pump)

If (say) the pressurized oil required from the Accumulator (working pressure for the actuator) is 60 bar, then,

Power in KW = PQ/600

(In American system it is PQ/1714 where P is in psi and Q is in GPM and output in hp)

Power in KW = (60 × 45) ÷ 600 = 4.5 KW = 5.85 hp ≈ 7.5 hp (Availability of motor ratings)

In case, it is decided not to have the accumulator, instead to go in for a larger capacity of pump and the motor the calculations are as follows.

Q = 15 l/min. during 5 seconds of the cycle,

Q = [15 l/min] ÷ 5/60 = 180 l/min.

Power in KW = (60 × 180) ÷ 600 = 18 KW = 23.4 hp ≈ 30HP (Considering motor ratings availability)

From the above we see that in the absence of an Accumulator, we would need almost 4 times bigger pump and the motor for the above referred example.

16.6.1 Accumulator unloading valve

We need to understand the working of this valve for including it in the circuits with Accumulators. This valve is similar to the system relief valve with only change being of the pilot line connection and incorporating a check valve between the pump line and the accumulator line.

Please refer the circuit in Figure 16 8.

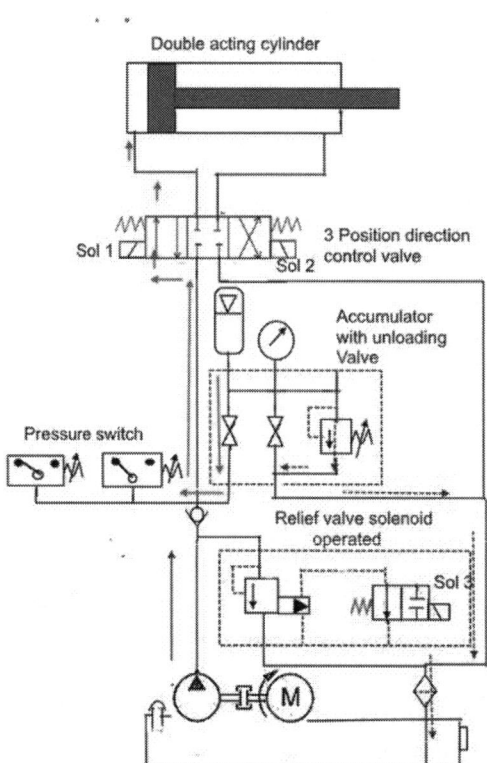

Figure 16 8 Accumulator circuit

There are two additions to this circuit.

1. Accumulators and its supporting elements- They are the accumulator, unloading relief valves and the pressure switches.

After assessing the demand for oil at different pressures, the pressure is set for opening (cut in pressure) the accumulator supply at a certain pressure when the oil flow is required by the actuator and the cut off pressure, when the oil supply from the Accumulator is to be cut off.

The Accumulator unloader valve ensures that the pressure inside the Accumulator is also set at a maximum designed pressure and if the pressure inside the Accumulator exceeds the limit, the Accumulator unloading valve relives the excess pressure oil to the tank.

The red line path shows the cylinder getting additional flow(during time t2) from the Accumulator when Sol 1 is energized so that the cylinder forward stroke is increased.

The yellow line is the average flow rate from the pump. (during time t1).

The blue lines indicate the flow through the unloader valve when the pressure inside the Accumulator exceeds the Accumulator set pressure.

The circuit also shows a system relief valve operated by a solenoid valve. In the type used in Figure 16 8, the relief valve shows the pump unloading the flow ,(when the pump is running), in the normal condition. If Sol 3 is energized, this unloading by relief valve path is blocked and the pump supply will be directed towards the circuit operation.

This unloading relief valve is added to this Accumulator circuit just as an example of the type of relief valve (solenoid operated) available. It is possible to add a standard system relief valve, even in the Accumulator circuit shown.

16.7 FEW OTHER APPLICATIONS OF ACCUMULATOR

1. Meeting emergency requirement of pressurized oil Compensation of flow during emergency breakdowns-In an emergency situation like unexpected power failure, the pressurized oil in the accumulator helps to out one or more output and/or return strokes.

2. System leakage-The compression force exerted by a hydraulic cylinder can only be maintained by compensating the inevitable losses due to system leakage. This is especially true in case of clamping operation where, any leakage could make the job to get loose of the clamp.. The accumulators are particularly suitable for this purpose of compensating for pressure loss during the operation.

3. Flow rate in pump output. -Pumps produce fluctuations in its flow rate and this could cause vibrations and damage to the hydraulic components. An Accumulator located close to the pump could take care of this eventuality. 4. Damping of pressure waves- In most of the hydraulic plants,

pressure waves are generated by various components or by the effect of load changes in the system, for example when using the bucket of an excavator. The installation of an accumulator protects the sensitive components from pressure waves.

16.8 POINTS TO RECALL

1. Hydro Pneumatic Accumulators are widely used in many applications.

2. The Accumulator needs following additional components – Prefill valve for charging the inert gas in to the accumulator.

Unloading valve for protecting the accumulator from excess pressure beyond its designed pressure.

17. HYDRAULIC CIRCUITS IN STACKABLE FORM

17.1 LEARNING OUTCOMES

By the end of this chapter the student will understand,

1. The hydraulic circuits drawn with stackable valves.

2. The different interfaces and manifolds of the components.

17.2 STACKABLE VALVES.

We have seen in the last chapter an example of system relief valve controlled by adding a direction control valve. Please refer Figure 17 1.

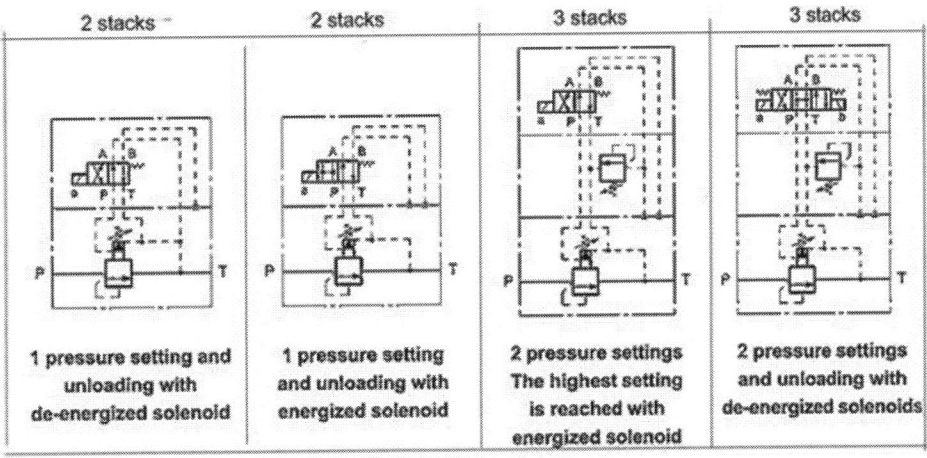

Figure 17 1 Stackable relief valves

The above is only to illustrate that the hydraulic valves can be connected by stacking one over the other.

The student by now must be able to read and understand a hydraulic circuit drawn in a conventional manner. (All the hydraulic circuit shown till now are drawn in the conventional manner.) However, if the piping between these valves are connected as per the conventional hydraulic circuits, then, it would require large length of piping. Further in a circuit having large components, if connected, in the conventional style, then, it would also be very confusing for maintenance.

The stackable valve present a neat alternative to this issue and present a clean appearance in addition to ease of maintenance. Each valve comes with an interface that accepts a valve below it and another one above it. Direction control valves are normally mounted as the top most element in the rack and hence would not require the top interface. The bottom interface would accept the valves below. Each of the valve at its bottom interface have a P line and a T line connection as well as connections to the two ports of the cylinder (Referred as A and B in the figures that follow.) As the valves come in different sizes for handing different flow rates, they are designated suitably and are commonly referred as modular valves.

Figure 17 2 Stackable Valves(Ref: Yuken India catalogue)

In Figure 17 2, the modular valves stacking is shown as an example. The physically stacked valves is shown Figure 17 3

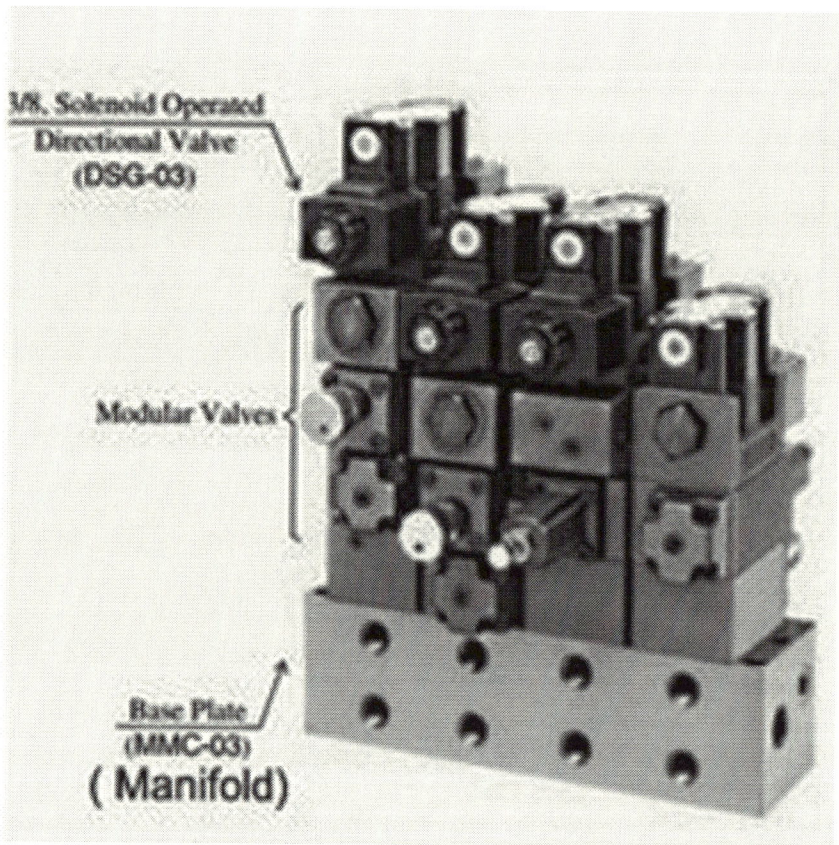

Figure 17 3 Stacked Valves (Courtesy Yuken India ltd.,)

Let us convert a simple conventionally drawn circuit to a modular valve version.

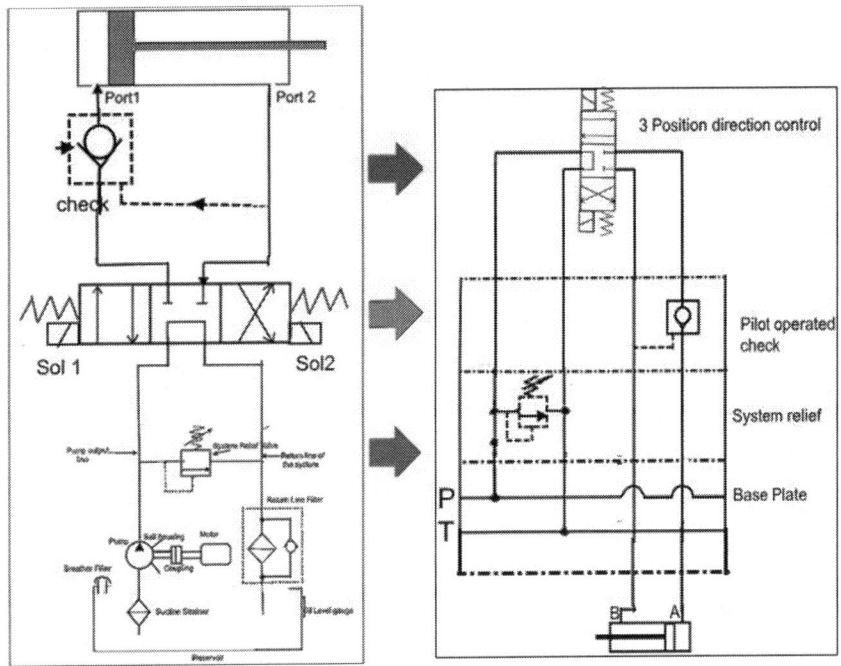

Figure 17 4 Hydraulic circuit drawn in conventional style Figure 17 5 Hydraulic circuit drawn with modular valves

Figure 17 4 and Figure 17 5 shows a simple circuit being drawn in both ways. At the time hydraulic design is being worked out a conscious decision is to be taken to go in for modular valves. Selection of modular valve sizes depend on the flow rate and the pressure ratings which are available with the hydraulic equipment manufacturers- The sizes an the common interface according to one of the manufacturers, Yuken India ltd., is given below.

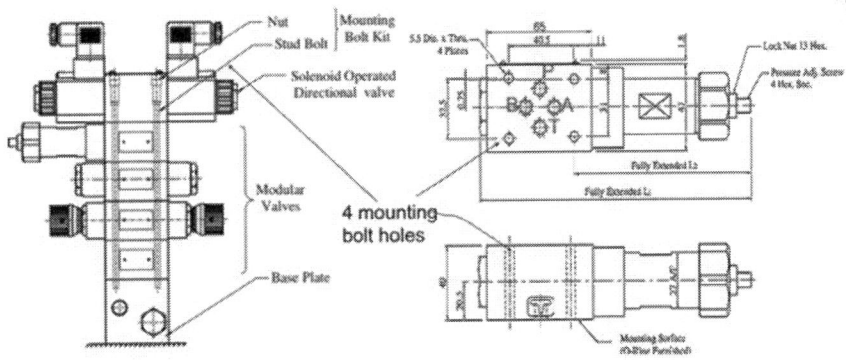

Figure 17 6 Size 01 series modular valves and its interface

As per one of the manufacturers'(Yuken India ltd.,) the stacking of modular valves and the interface for the starting range of valves are given in Figure 17 6.

Please note the P, T and A and B (A, B are actuator ports) marked red on the interface of this valve (It is a relief valve).

Similar interface drawings are available for bigger sizes of valves from any reputed hydraulic manufacturer.

Manufacturers data sheets also provide the order of stacking different functional valves. Generally the base plate for mounting the valves, o rings for mounting them and the bolt kits are also available from the hydraulic valve manufacturers.

17.3 POINTS TO RECALL

1. Designing the hydraulic system with modular valves help in simplifying hydraulic circuit connections and later in maintenance of the hydraulic power pack.

2. The order of stacking the hydraulic elements is similar to the conventional circuits, but the details can be obtained from the manufacturers' data sheets.

3. The student should be able to understand the circuit drawn either in conventional or in modular form.

18. INTRODUCTION TO PROPORTIONAL VALVES

18.1 LEARNING OUTCOMES

By the end of this chapter the student should understand

1. The functions of different proportional valves.

2. The hydraulic circuit using proportional valves.

18.2 PROPORTIONAL CONTROL VALVES

We have seen that, generally, the hydraulic valves can be classified under 3 categories – direction, flow and pressure control valves. Changing direction, flow or pressure during machine operation with these valves would require a separate individual valve for each -direction, flow or pressure desired.

Generally these conventional valves are also referred as 'switching' valves. In a conventional direction control valve, the solenoid is either on or off- The solenoids shifts the spool to either open or close positions only.

With proportional valves, it is possible to control infinite positioning of the spools, and, thereby, achieving infinitely adjustable flow volumes by using' Proportional' solenoids.

18.3 ADVANTAGES OF PROPORTIONAL VALVES COMPARED TO SWITCHING VALVES.

1. Adjustability of Valves	Infinitely adjustable flow and pressure via electrical input signal
	Automatic adjustment of flow and pressure during operation of the system.
2. Effect on the drives	Automatic and accurate adjustment of Force or torque Acceleration Velocity or speed Position of the piston or the rotor.
3. Energy	As pressure and flow are adjustable as per system requirement, energy is saved with proportional controls.
4. Circuit simplicity	less number of valves used in proportional hydraulics and hence circuit looks simplified.

Table 18 1 Advantages of proportional controls

Proportional valves are suited for applications

1. Requiring to vary either flow or pressure. The solenoids on these valves shift the spool according

to the voltage applied to proportional solenoids. in effect, they can change the speed at which the spool shifts or the distance that it travels. Because the spool in a proportional valve does not shift all the way, all at once, the valves can control the acceleration and deceleration of an actuator. Usually, varying shifting time of the spool controls acceleration and deceleration. Varying voltage to the solenoid coil limits spool travel to control the maximum speed of an actuator.

2. More precise actuator speed control is desired, compared to conventional valve controls.

The above two points are also clear from the Table 18 1.

18.4 PROPORTIONAL SOLENOIDS

Before we go on to proportional solenoids, let us understand the solenoids used in switching valves.

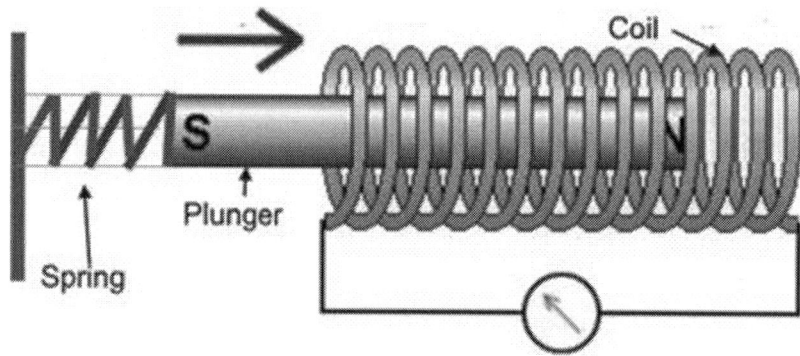

Figure 18 1 solenoid for switching coil

In Figure 18 1, the arrangement of a coil and a plunger with a spring is shown. The coil is energized by applying voltage across the coil. The current flows in the coil and creates a magnetic field. This magnetic field exerts a force on the plunger to move inside the coil against the spring force.

If the current through the coil is switched off, then the spring pulls back the plunger from inside the coil.

This plunger movement can be used to open or close a valve position like direction control valve. As you would understand the plunger position and hence the spool position of the direction control valve is either in activated position or in off position. The spool which is pushed by the plunger has no mid position or it is not possible to hold the position of the spool of the valve or the plunger in any finite position except in the two extreme positions.

In proportional solenoids, to put in simple language, it is possible to vary the current to the coil and thereby control the force exerted on the plunger. and thereby accurately position the plunger and hence the spool of the valve.

It is possible to calibrate the current to be proportionate to the force exerted on the armature / plunger and thereby the position of spool movement. The spool movement can be further accurately positioned by including LVDT (Linear Variable Differential Transformer) in the electrical circuit.

18.5 PROPORTIONAL CONTROL VALVES

The different proportional valves available and their symbols are given below.

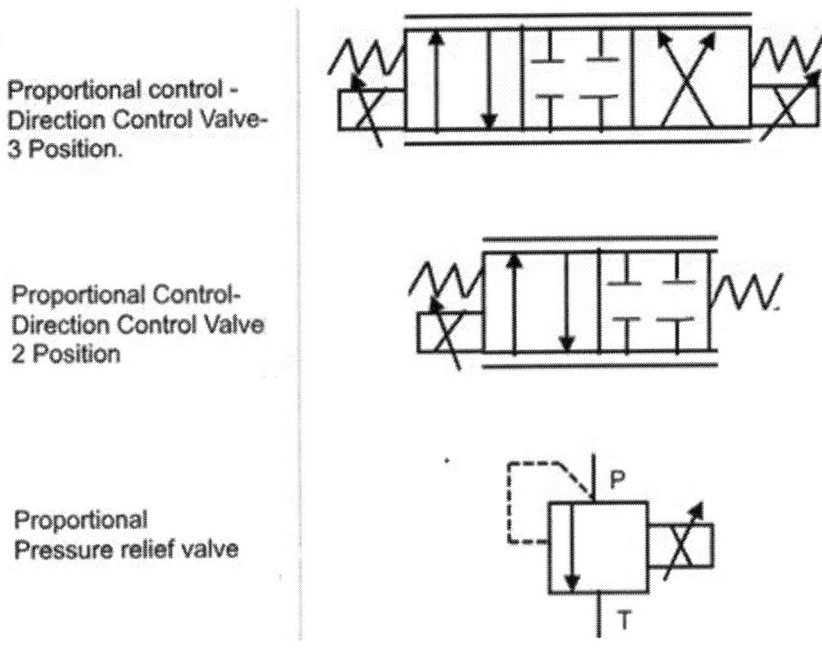

Figure 18 2 Symbols of popular proportional valves

All the valves shown in Figure 18 2 have identifications reserved for proportional valves. The solenoid of these valves are crossed with an arrow – meaning that the solenoid currents are adjustable.

The direction control valves also have two lines parallel to each other drawn above and below the positions of the valves.

We also need to know the process required for current flow variations to proportional solenoids.

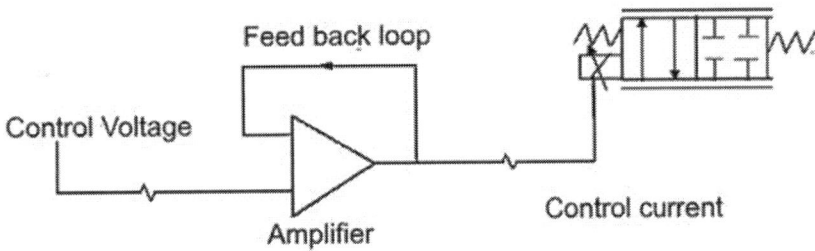

Figure 18 3 current flow in proportional solenoids

Please refer figure 18-3. In its simplest form, it shows the current control signals to the proportional solenoid.

1. A DC voltage generator acts upon an electrical amplifier.
2. The amplifier converts the voltage (input signal) into a current signal (out put signal)
3. The current acts on the proportional solenoid.

4. The proportional solenoid acts on the spool of the solenoid in proportion to the current signal.

5. The oil flow to the actuator is in proportion to the spool restriction of flow.

All the proportional valve manufacturers should be in a position to arrange for supply of electrical accessories required along with the proportional solenoid valves.

The hydraulic elements manufacturers are also in a position to offer pilot operated proportional control valves for larger flow rates required for some applications.

A simple hydraulic cylinder drive circuit with standard valves can be compared with that using proportional valves.

In the hydraulic circuit with proportional valves you would see that flow control valve is no longer needed for adjusting the speed of the actuator, when, proportional direction control valve is used.

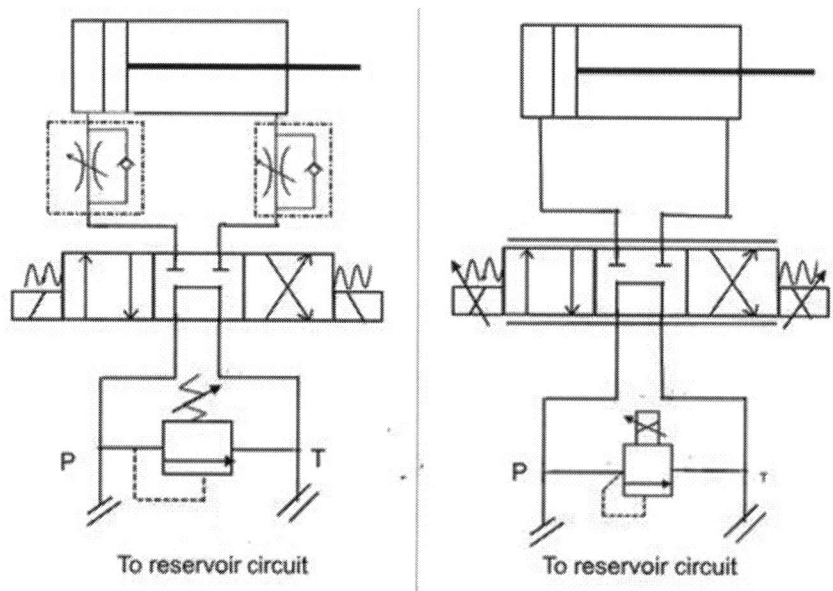

Figure 18 4 Hydraulic circuits with solenoid and proportional valves

Please refer Figure 18 4. In the figure on the circuit on the left, a flow control valve is used with standard solenoid operated 3 position direction control valve and a standard relief valve. On the right is the hydraulic circuit with proportional solenoid direction control as well as relief valve.

In the circuit on the left, usage of flow control valve is not shown as the same is not required since the proportional direction control valve itself operates restricting the oil flow into the actuator ports. Controlling the electrical signal current would influence.

A second control signal current acts on the proportional relief valve. The pressure can be continuously adjusted by means of this control signal. This facility of adjustment to helps to lower the pressure during reduced load phases.

18.6 POINTS TO RECALL

1. The benefits of using proportional hydraulics as against standard solenoid valves.
2. The electrical accessories required for activating proportional solenoids.

19. SERVO HYDRAULICS

19.1 LEARNING OUTCOMES

By the end of this chapter, the student will understand

1. The difference between proportional hydraulic valves and sero valves.
2. Applications where Servo hydraulic valves can be used.

19.2 WHAT IS SERVO HYDRAULICS

In terms of precision of movement of the actuator and fluid control, the starting point is direct solenoid operated valves. Then proportional valves and then finally the most precise of all three is Servo valves. The circuit /system having Servo valves can be referred as Servo hydraulic systems. In terms of cost, proportional valves bridge the gap between solenoid valves (least expensive) and the Servo valves(Most expensive).

19.3 COMPARISON OF PROPORTIONAL HYDRAULICS WITH SERVO HYDRAULICS

1. The need for filtration levels of pressurized fluid in the case of servo valve is much higher even compared to the levels required for proportional valves. Please refer Table 19 1

2. Proportional valve system normally do not need measuring systems- whereas Servo systems are incorporated with measuring systems and hence feedback loops are added- thereby making the system more complex and expensive compared to Proportional hydraulic systems.

3. Like in proportional valves, a low voltage is used to control the servo valve. The control voltage is passed into an amplifier which provides the power to alter the valve's position. The valve will then deliver a measured amount of fluid power to an actuator. The use of a feedback transducer on the actuator returns an electrical signal to the amplifier to correct and for precise measurement of the actuator.

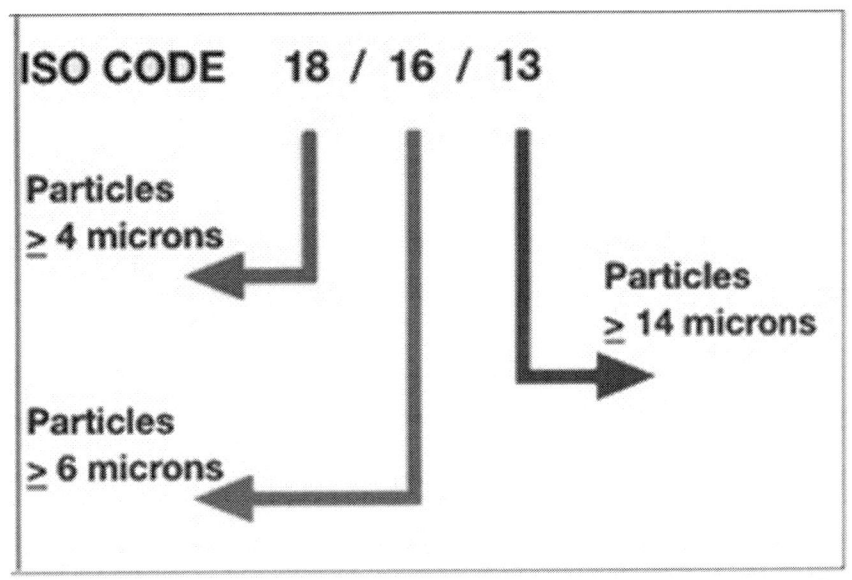

ISO Classification & Definition

Range number	Micron	Actual Particle Count Range (per ml)
18	4+	1300-2500
16	6+	320-640
13	14+	40-80

Fluid Cleanliness Required for Typical Hydraulic Components

Components	ISO Code
Servo Valves	16/14/11
Proportional Valves	17/15/12
Piston pumps and Motors	18/16/13
Directional/ Pressure controls	18/16/13
Gear pumps/ Motors	19/17/14
Flow control valves, Cylinders	20/18/15
New unused fluid	20/18/15

Table 19 1 Filtration level of hydraulic elements

From Table 19 1 Filtration level of hydraulic elements, you would note that the level of filtration required is very tight and hence proper care has to be taken before commissioning the system.

19.4 ELECTRO HYDRAULIC SERVO VALVES

These valves mostly find application in closed-loop hydraulic control systems- the applications are generally those in which very high performance was required- like in Aero space, defense and in high precision machine tool and in plastic molding industry.

Servomechanism is a control system which measures its own output and forces the output to quickly and accurately follow a set value by the feedback signal received through a transducer.

19.4.1 Servo Valves construction

We can consider three types of models in servo valves construction.

Single-stage- Direct operated: Torque motor supplies enough torque to shift the spool against the pressure.

Two-stage: Torque motor shifts the pilot spool (first stage), which directs flow to shift the second stage. The second stage supplies flow to the actuator.

Three-stage: Pilot stage shifts second stage, which shifts the third stage. Three-stage valves are used for application with high flow and high pressure.

We shall look at a single stage construction of servo valve.

The process of servo valve working starts with a signal current being generated by a Controller (Computer/ PLC/ Micro controller) and the current signal is sent to a torque motor through a current amplifier.

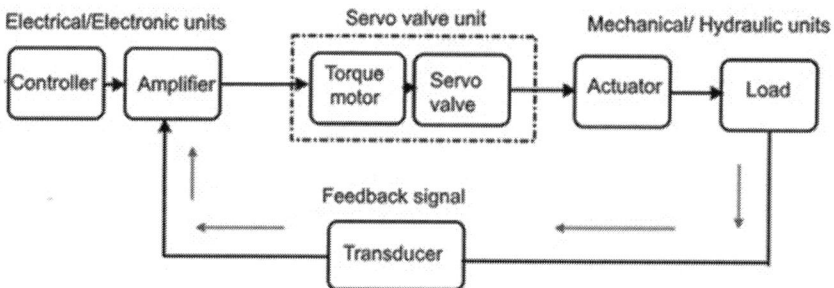

Figure 19 1 Servo valve system

The torque motor is directly connected to the spool of the valve. Depending on the movement of the torque motor, the spool position of the valve also get adjusted. If we change the current magnitude, the position of the torque motor and hence the spool would also change. The feedback current signal to the amplifier (summing amplifier) also comes from the transducer connected to the load. This signal is added or subtracted to make the spool location to be precisely at a point to deliver by the spool adjustment the measured flow/pressure of oil.

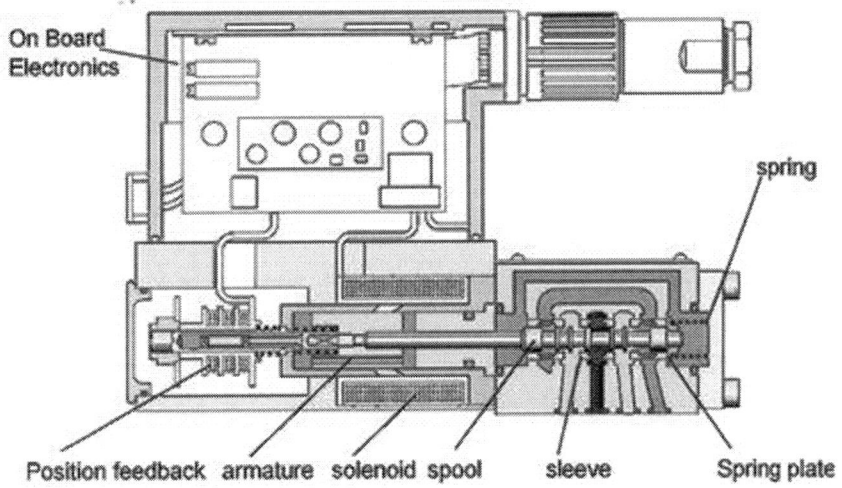

Figure 19 2 Direct operated servo valve (Rexroth)

In Figure 19 2, only the servo valve unit (Torque motor and the spool valve) with position feedback transducer (LVDT) are shown. It is to be noted here that the transducer is attached to the solenoid spool extension itself. In Figure 19 1, it is shown to be attached to the load. The direct operated servo valves are available generally from 2 l/min to 100l/min and for pressures less than 100 bar. If the flow rate exceeds 100 l/min, the option is to go in for two stage or 3 stage pilot operated servo valves.

19.5 POINTS TO RECALL

1. Servo valves need higher degree of contamination controls needing better filtration units in the hydraulic system.

2. The servo valves are high performance valves and are used when very precise performance parameters are enforced – like precise control of actuator positioning, pressure and flow rate maintenance.

20. LOGIC OR CARTRIDGE VALVES

20.1 LEARNING OUTCOMES

By the end of this chapter, the student will understand

1. The functions of cartridge valves and its constructional features
2. hydraulic circuits with cartridge valves.

20.2 WHAT IS A LOGIC/CARTRIDGE VALVE?

In chapter 17, we have seen stackable valves. The stackable valves use a base plate / sub plate/manifold on which the first deck of a modular valve is stacked.

In the same manner, we use a manifold with cavities to accommodate these cartridge valves. These cartridge valves offer a design alternative to conventional sliding spool valves. Hydraulic design engineers sometimes employ both cartridge and conventional sliding spool valves in one system. The cartridge valve can be designed as a screw in type (into the manifold cavity) or as a slip in type. Generally, manufacturers make screw in cartridge valves for less flow rates- (say) up to 120 l/min and slip in cartridges for minimum 200l/min and above.It is recommended that the user/ hydraulic designer finds out the specifications og cartridge valves while designing the circuit.

20.3 ADVANTAGES OF CARTRIDGE VALVES

The main reason for using slip-in cartridge valves in high-flow circuits at a reasonable cost. Large spool valves are available with high flow capacity but few are manufactured, making them expensive with long delivery times.

Another advantage of cartridge valve is its fast response on opening. Because there is no overlap, flow is almost instantaneous after the valve receives a start signal. Response also is fast on closing because the poppet only opens far enough to pass the flow going through it. This means it does not have to move any extra distance to start to restrict flow and shut it off.

Another factor is its size compared to its traditional counter parts – sliding spool valves- rated for a given range of flow and pressure.

20.4 CONSTRUCTIONAL FEATURES OF POPPET TYPE SLIP IN CARTRIDGE VALVES

Please refer Figure 20 1.

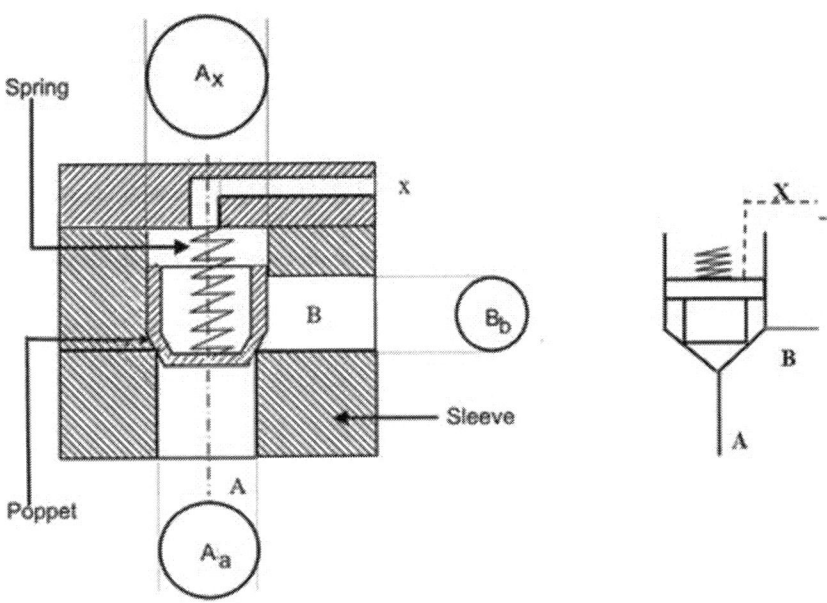

Figure 20 1 cartridge valve construction and its symbol

Cartridge valves/logic valves, have two operational ports A and B. The flow path between these two connections is controlled hydraulically by a pilot pressure applied to X. The cartridge valve includes a poppet which is held in the closed position by a spring. The poppet valve has a seated cone, giving a near zero leakage (dependent upon pilot control) condition across the two ports. The closing spring is retained by the control cover which encloses the cartridge valve and provides pilot connections from the X port. Various types of pilot control can be mounted either to the control cover or to an adjacent manifold face, to provide direct control of the cartridge valve.

The effective areas of the basic element are Aa ,Bb and Ax. Normally, the valves are manufactured in three sizes depending on the area ratios.

Ax,is effective area of the poppet; Aa and Bb are effective areas of port A and B respectively.

Area ratio 1:1 means area Ax and area Aa are equal. Bb = 0 and if the pilot line is connected to port B, then, the valve works like a check valve allowing oil to flow from port A to B and not from port B to A.

Area ratio 1:1.1 means area Ax is 1.1times area Aa.and, Bb =0.1 Aa

Area ratio 1:2 means Area Ax is two times that of area Aa and Bb = Aa

The oil pressure acts on these three different areas mentioned. The forces that hold a valve closed are the pressures acting on Ax plus the spring force. The forces that act to open the valve are the forces acting on port area Aa and Bb. If the sum of the closing forces is greater than the sum of the opening forces, then, the poppet remains closed. If the sum of opening forces is more than the closing forces, then, the valve gets opened.

At times, when, the valve is with ratio Ax = Aa, then it also referred as balanced poppet type of cartridge valve. In the other two cases when the areas are not equal (i.e) area Ax is greater than Aa, the valve type is also referred as unbalanced poppet type cartridge valve.

The valves come with different ratios as mentioned and also with different soring ratings.

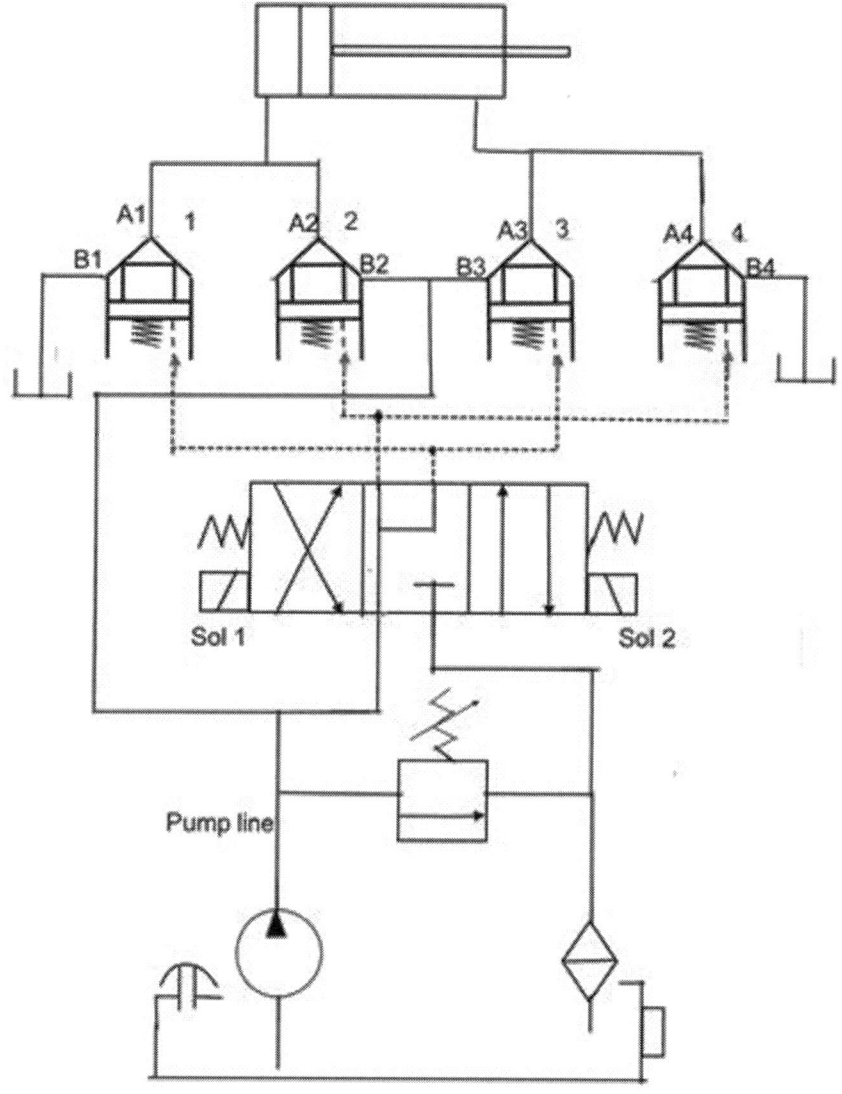

Figure 20 2 Hydraulic circuit with cartridge valves

Figure 20 2, shows how a large double-acting cylinder can be controlled using cartridge valves. A double-solenoid-operated direction control valve, with low flow rate,(costs less) feeds pilot pressure signals to four cartridge valves, marked 1,2.3,and 4.These valves are coupled in pairs to each end of the double-acting cylinder. The A and B ports of the valves are marked as A1, B1 etc., respectively for each cartridge valve.

One cartridge valve from each pair is permanently connected to the tank drain line (1 and 4-B1 and

B4 connected to the tank line) and the others (2 and 3- B2 and B3) to the pump pressure line. In the position drawn, all the four valves are held close by the pilot pressure signals, as they are all connected to the pump line in the neutral position of the solenoid valve. In the Figure 20 2, shown as little red lines. The cylinder position is locked.

When solenoid 1 is energized, pilot pressure is maintained on valves 1 and 3 which remain closed. But valves 2 and 4 are released to tank line.by the cross port passage of sol 1.The fluid under pressure, flows from the pump through the cartridge valve 2(B2 to A2 to cylinder port) to the full bore piston side of the cylinder and the cylinder extends. (Please note B3 is also connected to pump line, but, as the valve is kept closed, A3 is not open)The fluid from the rod end of the cylinder flows through valve 4 back to the tank.

When solenoid 2 is energized, cartridge valves 2 and 4 are closed under pilot pressure and valves 1 and 3 are released. Pump line connected to cartridge valve 3(B3 to A3 to the rod side port of cylinder) causing the cylinder to retract. The oil from the full bore side of the piston port finding its way to the tank through valve 1.

Therefore, with a combination of a small conventional 3 position direction control valve and four cartridge valves, it is possible to handle higher flow rates and actuate a bigger sized hydraulic actuator.

20.5 POINTS TO RECALL

1. Cartridge valves offer us an alternative especially when large flow rates are to be handled. While designing a circuit with cartridge valves,a few conventional valves are also used in the circuit.
2. Selection of cartridge valves include connecting passage areas as well as the spring pressure.

21. APPENDIX 1 - HYDRAULIC TERMINOLOGY

Annular Area

Refers to the area on the piston rod side area of a cylinder.

Back pressure

Back pressure occurs when hydraulic flow in the return line is restricted and causes a buildup of pressure backward through the line. This is undesirable as it robs the entire system of potential flow as the pump now has to produce more power to overcome the back pressure. Hydraulic motors in particular cannot handle excessive back pressure and will prematurely fail if exposed for any length of time.

Baffle

Normally refers to a plate installed inside the reservoir to separate pump inlet from delivery side.

Bleed off

To divert a part of pump flow to tank line.

CETOP

CETOP is the abbreviation of Comité European des Transmissions Oléohydrauliques et Pneumatiques. CETOP is the European Fluid Power Committee. Thus, it is: · The communication platform for fluid power in Europe.

Closed center valve

A valve(generally a direction control valve)in which all ports are closed or blocked in neutral position.

Closed loop

A system in which the feedback is compared to the set value and correction is applied to control the output of the loop.

Compensator control

A displacement control for vatable pumps/ motors to alter displacement when system pressure exceeds its adjusted pressure setting.

Contamination

Foreign particles present in the hydraulic medium(oil) adversely affecting system performance. These can be present in solid/liquid or gaseous forms.

Counterbalance valve

A pressure control valve which maintains back pressure to prevent free falling loads.

Cracking Pressure

Pressure at which a pressure actuated valve starts to open to pass the pressurized fluid.

Differential cylinder

Any cylinder in which the areas on either side of the piston are not equal.

Displacement

The quantity of fluid(oil) from a pump/ motor for one revolution or in one stroke of the cylinder.

Double acting cylinder

Hydraulic medium (oil) is used for both forward and return stroke of the cylinder.

Drain(External/Internal)

A separate passage meant to discharge leakage fluid(oil) to the tank(Reservoir).Often in hydraulic systems, hydraulic motors and pumps run case drain hoses. The reason for this is to drain excess internal oil leakage from the motor. This is referred as external drain. In the case of internal drain the leakage from the equipment goes inside the equipment itself and then the leaked fluid finds itself to the tank line or used by the equipment itself.

If this drain provision is not given, it would result in damage to the seals.

Feedback

The output signal from a feedback device

Flooded

A condition where the pump inlet is positively charged by placing the reservoir oil level above the pump inlet port.

Fluid

Liquid or gas.

Frequency

Number of times, the action occurs in unit time.

Head

The height or a column of fluid above a particular line and expressed in linear units. Pressure equals the height times the density of the fluid.

Hydrodynamics

Engineering science relating to the liquid flow and pressure.

Hydrostatics

Engineering science relating to the energy of confined liquids

Kinetic energy

Energy that a substance has by virtue of its mass and velocity.

Laminar Flow

A condition where the fluid particles move in continuous parallel paths.

Manifold

A Machined part which provides multiple connection/ passage ports.

Manual Override

A method / means by which a manual actuation is possible in an automated device/

Micron

One millionth of a meter or about 0.00004 inch.

Micron rating

The smallest size particle a filter will remove.

Open center valve

A valve(generally a direction control valve)in which all ports are in neutral position.

Pilot Pressure

Auxiliary pressure used to actuate or control a hydraulic element or component.

Pilot valve

An auxiliary valve used to control the operation of another valve.

Poppet

A valve that moves perpendicular to or from its seat.

Positive displacement

A characteristic of a pump or motor where the delivery(outlet side) is sealed from the suction (inlet side). They do not mix as in centrifugal or other types.

Pre-charge Pressure

The pressure of compressed gas inside the accumulator prior to admission of Fluid.

Pressure drop

The difference in pressure between any two pints in a system.

Pressure line

The line carrying the oil(fluid) from the pump outlet.

Pressure override

The difference between the cracking pressure(of a valve)and the pressure at which the valve is working at its full rated flow.

Ram

A single acting cylinder with a single diameter plunger instead of apiston and a rod.

Regenerative circuit

A piping arrangement for a differential type cylinder in which discharge fluid from the rod side combine with pump delivery to be directed to the head end.

Return line

A line that carries exhaust fluid(oil) from the actuator to the tank(Reservoir).

Single acting cylinder

A cylinder in which fluid(oil) is used for only one movement (Forward/ return). The other motion is by spring or by gravity.

Spool

A cylindrically shaped part of a valve which is moved to open/close passages for the fluid(oil) to flow.

Sub plate

A mounting plate for valves and that provides for connection to piping.

Suction line

The hydraulic line connecting the pump inlet port to the reservoir.

Tank

Reservoir.

Torque motor

An electromechanical transducer having rotary motion to actuate servo valves.

Transducer

A device that converts one type of energy to another. An example would be that of a device that coverts a displacement in to electrical signal(LVDT).

Turbulent flow

A condition where the fluid particles move in random paths rather than in parallel paths as in laminar flow.

Vent

Normally refers to an opening (in a valve or a device)to atmospheric pressure.

22. APPENDIX2-USEFUL FORMULAE(FPS)

1. Flow velocity

Suction Line Low Pressure 2-4 Feet/Second

Return Lines Low Pressure 10-15 Feet/Second

Medium Pressure Lines 500-3,000 PSI (30 to 150 bar) 15-20 Feet/Second.

High Pressure Lines above 3,000 PSI (150 bar) 25-30 Feet/Second

2. Hydraulic motors

Torque (lb. in.) = (HP X 63,025)/2 π

For lb-ft use 5,252 Constant in Place of 63,025

Torque (lb. in.) = [PSI x Displacement (in3/Revolution)]/2

Torque (lb. in.) = [GPM x PSI x 36.77]/RPM

Torque (lb. in.) = [Motor Displacement (in3/Revolution)]/0.0628

Horsepower = [Torque (lb.-in) x RPM]/83025

Flow Rate at 100% Efficiency: Q (Flow rate in GPM)

=Efficiency X [RPM X In3 per revolution]/231x (in3 per gallon)

Your feedbacks are welcome- ilango.sivaraman@gmail.com

23. APPENDIX3-BIBILIOGRAPHY

Kerstin Avila1, D. M. (2011). The Onset of Turbulence in Pipe Flow. Science, 192-196.

pd Vs centri.pdf. (2007). Retrieved from www.pumpschool.com:

http://www.pumpschool.com/intro/pd%20vs%20centrif.pdf

www.yukenindia.com

Fluid Power Circuits and Controls, John S. Cundiff, 2001

nptel.ac.in/courses/112106175/Module%202/Lecture%2019.pdf

hydraulicspneumatics.com › Technologies › Other Technologies

13th International Conference on AEROSPACE SCIENCES & AVIATION TECHNOLOGY, ASAT- 13

www.moog.com/literature/ICD/Moog-CartridgeValves-2way_Stard-Catalog-en.pdf

https://www.parker.com/literature/.../HY11.../HY11-3500UK_10.2011_PDFoverall.pdf

Linköpings universitet TMHP51 IEI / Fluid and Mechanical Engineering Systems

https://www.boschrexroth.com

Made in the USA
Lexington, KY
18 August 2018